Benjamin Burkhart
Platform Jazz

Popular Music

Benjamin Burkhart (Dr.) is a musicologist at the Hanover University of Music, Drama and Media. His research and teaching focus on the media cultures, analysis, and social history of popular music and jazz. Prior to his work in Hanover, he served as an interim professor for the history and theory of popular music at the University of Giessen, a senior scientist at the Institute for Jazz Research at the University of Music and Performing Arts Graz, and a research associate at the Center for Popular Culture and Music at the University of Freiburg.

Benjamin Burkhart

Platform Jazz

Algorithmic Music Culture on TikTok

[transcript]

This publication received financial support from the publication fund NiedersachsenOPEN, funded by zukunft.niedersachsen

Bibliographic information published by the Deutsche Nationalbibliothek

The Deutsche Nationalbibliothek lists this publication in the Deutsche Nationalbibliografie; detailed bibliographic data are available in the Internet at https://dnb.dnb.de

transcript Verlag | Hermannstraße 26 | D-33602 Bielefeld | live@transcript-verlag.de

Cover design: Kordula Röckenhaus
Printing: Elanders Waiblingen GmbH, Waiblingen
Proofread by Nicola Heine & Benjamin Burkhart
https://doi.org/10.14361/9783839431597
Print-ISBN: 978-3-8376-7913-7 | PDF-ISBN: 978-3-8394-3159-7
ISSN of series: 2747-3066 | eISSN of series: 2747-3074

Printed on permanent acid-free text paper.

Contents

1. Introduction

“Jazz on TikTok? How’s that supposed to work?”

“TikTok, isn’t that that silly platform for teenagers? They’re going to play jazz on it?”

“What? There’s jazz on TikTok?”

“That doesn’t go together at all!”

Astonishment, skepticism, disbelief, or simply bemused giggles. These are the kinds of reactions that I have been encountering in conversations with colleagues, students, and in my personal circles since beginning my in-depth research on the topic of jazz on TikTok in early 2022. Many people apparently do not expect this type of music to play a role on this platform at all. But why is that?

TikTok is a short-form video platform that has existed in its current form since 2018. It is operated by the Chinese technology company ByteDance and emerged from the platform Musical.ly, which was launched in 2014. The term *short-form video* initially referred to the policy of allowing videos of up to fifteen seconds in length to be uploaded to TikTok. While it is now both possible and common practice to upload longer videos, the platform is centered around videos of fifteen to sixty seconds in length. TikTok experienced a rapid ascent to popularity following its initial release, with a notable acceleration during the global pandemic. During the months of the first global lockdowns in 2020, the platform witnessed a surge in both downloads and user numbers, exceeding previous records.

At the outset, lip-syncing videos and dance challenges constituted the predominant content on TikTok. However, the platform soon began to accommodate a more diverse array of content. At present, a variety of content is available on the platform, including videos focused on climate activism and science communication. Even politicians and political parties are utilizing TikTok as a communication tool. The platform's content is diverse, yet it is still perceived by some as a platform for young people with a focus on silly content.

Reusing sounds is a significant aspect of the TikTok platform. This refers to the use of segments taken from musical compositions, such as part of a melody, a vocal line, or a chorus, in the creation of TikTok videos. Such excerpts may then be utilized by other users as a foundation for new videos, thereby facilitating their dissemination. TikTok is often the subject of criticism precisely because of this mimetic logic, which might conceivably, as the widely accepted conclusion suggests, lead to songs being produced *for* TikTok, so to speak. This implies a compositional or production-aesthetic tailoring along certain functional logics of the platform, with the aim of ensuring that certain song snippets *work better* on TikTok. In other words, they are more likely to enter circulation, thereby increasing the overall recognition of the songs.

The supposed emphasis on frivolous content, lip-syncing, dance challenges, and brief musical excerpts may initially appear to be incompatible with the conventions of jazz. How can improvisations lasting several minutes and intensifying in expression, complex harmonic structures or spontaneous ensemble interactions that emerge in the course of live performances be presented in a short-form video format? These elements undoubtedly play a crucial role within the realm of jazz. But what is actually presented as jazz on TikTok? Are there facets of jazz that can be transferred into a short-form video context with comparative ease? Are there certain jazz styles that might work particularly well in this media context? Isn't it obvious that jazz musicians respond to the technological innovations of their time, and – wishing to benefit from new communication media and the corresponding marketing potential – therefore also become active on new platforms like TikTok? And shouldn't it be the task of jazz research to take a closer look at such developments?

My research project on jazz on TikTok was inspired by my interest in exploring the ways in which jazz can be positioned within broader discourses surrounding technology. Having been academically socialized in popular music and jazz research, I had, among other things, worked for several years on a research project on playback devices in popular music cultures, with a focus on cultural and technical history. In the field of popular music research, the investigation of technologies, technical artifacts of various kinds, and human interactions with such objects is a key focus. This encompasses a range of topics, including playback devices, recording and storage media, musical instruments, effect devices, and the socio-technical constellations at play in recording studios and when working with digital audio workstations (DAWs). In the field of jazz research, these topics are given considerably less attention, despite the fact that the aforementioned technologies and the manner in which individuals interact with them play a significant role in the context of jazz as well.

At approximately the same time, research on digital platforms pertaining to music piqued my interest. The literature in this field repeatedly addresses the ways in which musicians, producers, and label employees navigate the demands placed upon them by digital platforms, as well as the extent to which the algorithms of these platforms influence the processes involved in the distribution, reception, and production of music. I subsequently began to observe the representation of jazz on various digital platforms, initially mainly on YouTube and Spotify. In doing so, I gained the impression that jazz can be presented on different platforms with very different focuses. On the one hand, this means that very specific facets of jazz achieve particular visibility on certain platforms. On the other hand, jazz can be integrated into media contexts on platforms that did not play a significant role in this musical culture before the major breakthrough of digital platforms from the late 2000s onwards. Such observations can be exemplified by the dissemination of so-called *coffee table jazz* or *chill jazz* via highly popular Spotify playlists (Burkhart 2022). The advent of platforms has also facilitated the dissemination of music videos by jazz musicians on YouTube. Prior to this, such videos were uncommon in the jazz genre. However, they are now regularly produced by jazz musicians and are also used, among other things, to articulate socially critical messages (Burkhart 2025).

Based on these observations, I decided to examine the representation of jazz on at least one more platform at the beginning of 2022. At that time, TikTok was already a highly popular platform, yet I was largely unfamiliar with it. However, given its sizeable user base and the clear differences to platforms like Spotify and YouTube, which are immediately apparent due to TikTok's focus on short-form video content, I believed it would be worthwhile to explore this platform further. After familiarizing myself with TikTok and looking for jazz musicians who are active and popular on the platform, I quickly formed some initial and already quite specific impressions. I came across names like Laufey, Ricky Rosen, Rachel Chiu and Stacey Ryan – the latter describes herself on her TikTok profile as the "queen of jazztok." I hadn't heard of any of these names before, but the young musicians (all in their twenties) already had a six-figure, sometimes even seven-figure following. In their videos, they performed classic jazz standards such as "It Don't Mean a Thing (If It Ain't Got that Swing)" and "Fly Me to the Moon." They often garnered considerable attention, with numerous videos amassing millions of views. My impression after initial observations was that specific facets of jazz also seemed to attract particular attention on TikTok, or perhaps even to be afforded by certain platform logics.

The central inquiry of this research project is to examine how jazz is represented on TikTok and to identify the representation of this musical culture that emerges on this platform. Does the idea of jazz that emerges on TikTok differ from those constructed in other media contexts, such as journalism, the book market, documentaries, or in jazz research? What role does the musicians' engagement with the functional logics of the platform play, especially with its algorithmic system? What happens to the representation of a music culture when it is transferred into a new media economic context in which, at least according to the often astonished reactions that my research project provoked, one would not expect it?

In order to reconstruct the specific circumstances underlying the representation of jazz on TikTok, this study centers the voices of musicians who are particularly successful on TikTok with jazz-related content. These individuals have experience in working with TikTok and offer first-hand accounts of popularization processes on the platform. Empirically retracing the steps involved in their platform-related actions promises valuable insights into the question of which jazz-related content is featured the most prominently on the platform and for what reasons. In particular, the focus is on processes of cultural hierarchization in the context of TikTok. We can assume that in the competition for visibility in digital spaces – especially in the context of a platform like TikTok, where audio-*visual* content is shared – social categories of difference such as *race* and gender, as well as specific body norms and the cultural power relations associated with them, also play a crucial role.

The following questions are at the center of the study: Which jazz styles are prioritized by popular musicians on TikTok and consequently become particularly visible on the platform, and why? Do specific facets of jazz appear to be more suitable for this platform and its short-form video format than others? What role do the algorithmic moderation logics of the platform play in the considerations of the musicians regarding the popularization of certain content? To what extent can musicians see through these logics and to what extent do they allow themselves to be guided by them? What, then, is the logic of the socio-technical interactions between the platform and the musicians? Do these interaction processes have a homogenizing function, in that certain content is prioritized, both algorithmically and by the musicians, or through the socio-technical interplay of human and non-human actors? And are familiar cultural hegemonies perpetuated in the process, or do shifts in emphasis with regard to cultural hierarchies occur in the context of the platform?

The study commences with an examination of prior research on the nexus of jazz and online media (chapter 2), with the objective of highlighting the extent to which this domain has been explored in the context of jazz research in recent years and the principal subjects that have been addressed. What is the role of research on digital platforms in this regard? And does this research illuminate any tangible approaches that the present study can build upon?

In order to evaluate the potential impact of digital platforms on culture, it is essential to comprehend their underlying operating principles. Chapter 3 presents an overview of the core modes of operation that are intrinsic to platforms, as well as an examination of core methodologies employed in platform studies. How is the data with which platforms operate collected, processed, and monetized? How do platforms control and curate their content? What is the basis for the action-structuring potential of platforms? And what research approaches can help to better understand the potential influence of platforms on processes of cultural production?

Subsequently, the specifics of the TikTok platform are outlined, particularly with regard to the platform-specific potential to influence processes of cultural production and the representation of cultures in digital spaces (chapter 4). The chapter provides an overview of the platform's central functionalities and of previous findings on the question of which TikTok-specific conventions have already emerged in recent years, i.e., which content has been more likely to be popularized on TikTok so far.

In chapter 5, my research approach is delineated. The chapter outlines the requisite methodological steps for empirically investigating the background of a particular representation of music cultures on TikTok. The progression from a systematizing corpus analysis to a detailed interview study is fundamental. In the context of a corpus-analytical approach, the first step is to identify which content is presented as jazz and achieves the greatest reach on TikTok. Subsequently, the most popular videos in this area are described (e.g., which musical repertoires and musicians exert the greatest influence within this segment?). The interview study focuses on the jazz musicians with the widest reach on TikTok in order to gain insights into their platform-specific production logics. The interviews are combined with the *conversational analysis* approach. In essence, this means that selected TikTok videos of the interviewees were viewed and analyzed together in order to understand which specific video elements are part of the musicians' platform-related strategies.

In chapter 6, the results of the interview study are presented in detail. Among other things, the following topics were discussed in dialog with the musicians: Who are the musicians who are particularly successful with jazz-related content on TikTok? What is their relationship to jazz? What constitutes a successful career on the TikTok platform? What is being done to achieve success and professionalize TikTok-related content creation? How do musicians interact with the platform's algorithms, and to what extent do they develop specific strategies for action regarding TikTok's algorithmic system? From their perspective, does TikTok's platform logic make it more likely that certain jazz-specific repertoires are played rather than others, and that certain musical expressions are more likely to become popular on the platform? Do certain groups of people find it easier to become popular on TikTok than others? Do musicians benefit from their success on TikTok beyond the platform? And to what extent could the TikTok-specific representation of jazz influence developments in the field of jazz beyond the platform?

Chapter 7 then presents an analysis of the extent to which processes of cultural hierarchization are reproduced on TikTok in the field of jazz. Which elements of the previously analyzed processes on TikTok are genuinely novel, and which are not? Does TikTok reflect cultural hierarchies that have been previously documented in comparable forms (not only) in music cultures over an extended period of time? Is it possible that shifts in emphasis may occur in digital spaces and as a result of the influence of non-human actors with regard to the hierarchization of cultural objects and groups of people?

The study concludes with a discussion of the challenges inherent in examining a contemporary and dynamic subject such as TikTok as part of a multi-year research project (chapter 8).

2. Jazz Research and Online Media

Research on jazz and online media is scarce. The first relevant studies on this topic were published in the late 2000s, but only a few isolated works have been added since then. Consequently, there has as yet been no comprehensive and systematic reflection on jazz-related production and reception practices in digital spaces. Some of the few studies that do exist argue that this is not the fault of jazz scholarship alone, but that jazz musicians have for a long time refused to engage with online media developments. As Haftor Medbøe and José Dias argued in the mid-2010s, "[j]azz has been slow to embrace the power of social media and seems to consistently arrive late at the table be it in the examples of MySpace, Facebook and Twitter" (Medbøe and Dias 2014, n.p.). The authors posit that the jazz sector has historically displayed a preference for maintaining the status quo and reverting to the familiar, evoking the *better times* preceding the advent of the internet. As Medbøe and Dias go on to argue, this skeptical attitude is particularly surprising given that jazz, throughout the many decades of its history, has been continuously and closely linked to innovations in media and technology. These include early talkies, radio, the phonograph record, and developments in recording studio technology that made new forms of sonic expression possible (Medbøe and Dias 2014, n.p.; cf. also Smudits 2007). However, Medbøe and Dias do not provide empirical data to substantiate their assertions on the reluctance of jazz musicians to embrace online media. Consequently, their arguments read more like personal assessments than concrete research findings. The extreme paucity of literature on the relationship between jazz and online media makes it clear that at least jazz research is still skeptical about this topic.

While Medbøe and Dias focus on jazz musicians, a number of researchers have studied the relationship between jazz fans and online media, focusing in particular on the formation of jazz-related communities in digital spaces. The studies in question were all published between the late 2000s and the mid-2010s, which sets a limitation on the conclusions that can be drawn about the current situation. In a 2010 article on "online jazz fandom activities" (Wall and Dubber 2010, 159), Tim Wall and Andrew Dubber found that jazz fans did not differ significantly from fans of other popular music cultures in terms of their music-related online activities. For example, blogs, forums, web radios, video portals, and early versions of streaming services were utilized to gain knowledge about jazz and listen to music (Wall and Dubber 2010, 161–62). Additionally, relevant research on jazz-related digital spaces has been published by Tom Sykes and Ken Prouty. Sykes (2017) examines the information portal London Jazz News and the Facebook group Gypsy jazz uk (see also Sykes 2013 for details), while Prouty (2012a; 2012b) focuses on the article on jazz published in the online encyclopedia Wikipedia and the forum of the website All About Jazz. Such research shows that online forums are a preferred medium for jazz enthusiasts, particularly for discussing live events (Wall and Dubber 2010, 162). Additionally, by the early 2010s, many jazz fans were already utilizing online portals to learn about and purchase recordings (Sykes 2012, 85). In the 2000s, specific fan and collector practices in the jazz sector also migrated to digital spaces. One such practice is the digitization of rare and hard-to-find jazz records by an online community. This practice, as described by Simon Barber in 2019, involves ripping records and sharing the digitized versions online (Barber 2019, 455). Simon Barber's study references the website My Jazz World, which was online from 2007 to 2010 and enjoyed great popularity among jazz fans seeking obscure jazz records. In short, jazz fans have long been aware of the possibilities of sharing, researching, and listening to music that the internet offers. According to Tom Sykes, online media can be utilized by jazz fans in conjunction with their existing practices, without rendering jazz-related activities obsolete (Sykes 2017, 54–55).

Some of the authors also inquire as to the extent to which the shift of jazz-related activities, particularly those of fans, into digital spaces could potentially influence established jazz discourses that are negotiated offline. While in the pre-internet era, individuals such as music journalists and institutions such as conservatories could be considered particularly powerful discursive authorities in the field of jazz – and certainly still can be to a certain extent – the question arises as to whether or to what extent fans or private individuals can influence jazz-related discourses, given that they are now able to disseminate their opinions and attitudes toward jazz in forums or comment sections of social media platforms, for example. Ken Prouty's work on virtual jazz communities, in particular, addresses a complex array of related issues (Prouty 2012a,

2012b). Prouty enquires into the processes of jazz-specific community-building and knowledge generation in digital spaces. In order to illustrate how digital spaces facilitate the formation of jazz-specific communities and the generation of knowledge, Prouty investigates Wikipedia and All About Jazz as case studies, in addition to Facebook and YouTube, which were already well-established at the time. He demonstrates that these contexts gave rise to their own forms of jazz-related debates, with the Wikipedia article on jazz, for instance, being collaboratively written and continually revised on the basis of sometimes heated discussions. According to Prouty, the primary distinction between public offline debates and those on Wikipedia is that private individuals, not just journalists and academics, have their say in the latter context (Prouty 2012b, 129).

In this context, there is a recurring argument that the balance of power in jazz-related discourses has shifted since the 2000s. Wall and Dubber also argue that online media have made it possible for jazz fans to disseminate their own narratives about jazz more easily, more regularly, and with greater reach than was previously possible, for example, through letters to the editor in professional journals (Wall and Dubber 2010, 162). However, according to Wall and Dubber, hierarchies also emerge in digital spaces, where there are also differences in access to knowledge and music. For example, rare and expensive records are only accessible to those who can afford them. In addition, administrators of forums and operators of internet portals for exchanging and buying records could also be regarded as agents of discourse and taste formation (Barber 2019, 455). Occasionally, researchers argue that the more traditional and potentially taste-forming entities in the jazz sector are increasingly losing their significance. For example, in his work on the production of jazz under the conditions of digitalization, Dean S. Reynolds states that:

> Not only do new media provide adequate alternatives to essential aspects of the traditional recording industry, but they often obviate the need for cultural "gatekeepers" at record labels, newspapers, and other institutions, so that more musicians can deliver more music to larger audiences more quickly and at a lower cost. (Reynolds 2017, 181)

Although Ken Prouty asserts that traditional gatekeepers are not simply becoming obsolete in the age of online media, he also posits that a "fundamental decentralization of jazz discourse" is occurring (Prouty 2012b, 150). While journalists and other authorities would not completely lose their influence, the core argument goes, the internet would certainly provide constructive spaces for a diversification of discourse participants.

A limited number of studies have examined the experiences of jazz musicians with online media. William C. Banfield asserts that jazz must expand its audience and that young jazz musicians must cultivate new markets (Banfield 2022, 98). This is particularly feasible and imperative in digital spaces, particularly given that there is for the most part no noticeable refusal to engage with online media among young musicians (Banfield 2022, 106). Reynolds also emphasizes, with reference to one of his interviews with jazz musicians, that "jazz musicians are starting to 'think digitally'" (Reynolds 2017, 157). Reynolds then goes on to state, as might be expected, that jazz musicians must now be able to access the internet if they wish to remain competitive. According to Reynolds, it is now common for jazz musicians to distribute their own music via platforms such as Bandcamp and SoundCloud. These two platforms in particular allow musicians to disseminate their own music and curate their own content (Reynolds 2017, 159–62). YouTube is of significant importance due to its audiovisual component, which is why many jazz musicians have begun to post performance videos or live recordings of studio sessions on the platform. The band Snarky Puppy, for example, has benefited to some extent from the use of such videos (Reynolds 2017, 163–65). As Reynolds demonstrates in his interviews, many jazz musicians now utilize digital media platforms for purposes beyond the mere publication of new music or videos. Rather, they "cultivate social profiles" (Reynolds 2017, 176). The practice of posting status updates and photos documenting everyday life on tour or in the studio or sharing Instagram stories has become a regular aspect of the daily business of many jazz musicians. It is therefore unsurprising that those who wish to survive in the music marketplace are not much different from their counterparts in other fields in this regard.

Some authors also reflect on the extent to which online media have a concrete impact on the performance of music or the creative practices of jazz musicians. Tom Sykes, for example, argues that new media can certainly create new relationships between users and media technologies and thus influence the production of cultural content (Sykes 2009, 309–10). It is assumed that certain media environments offer music creators specific frameworks that favor the production of certain specific content, which can only be produced in its particular form within these frameworks. Chris J. Cottell presents concrete results of such processes using the example of videos published by the musician Jacob Collier on Facebook and YouTube, as well as transcription videos shared by jazz musicians on Facebook (Cottell 2021). The Collier videos, in particular, demonstrate how certain media environments can facilitate or enable the production and distribution of certain specific content. In his article, Cottell focuses on Collier's *#IHarmU* videos, in which he edits videos sent to him by fans of themselves performing short vocal or instrumental parts. Collier adds various audio-visual tracks to these videos in a split-screen process, harmonizing with the source material. This has led to the development of a dynamic exchange between Collier and his fans, as well as a distinctive form of audiovisual and musical collaboration. Neither would be possible without the technological and media infrastructure provided by digital platforms such as YouTube and Facebook. Cottell describes the "influence of audiovisual content-focused social media platforms on artistic practice" (Cottell 2021, 3) and also argues that a "shift in social understandings of existing musical practices" (Cottell 2021, 3) is also taking place in the context of digital media platforms. There is no doubt that young musicians like Jacob Collier, who are well known for their openness to diverse musical repertoires and towards the possibilities of expanding creative practices in digital spaces, are triggering a new (actually very old) discussion of what jazz may or can be under changing media-technological conditions.

YouTube, in particular, offers a diverse array of audiovisual representations of jazz. As Prouty (2012b, 145–47) notes, the platform encompasses a range of jazz-related content, including live recordings, tribute videos by fans dedicated to renowned musicians, and instructional videos. It is also noteworthy that some jazz musicians are now utilizing a media format that has been employed primarily in pop music culture for decades: the creation of music videos. These videos not only document the music-making process, as is the case with filmed concert performances and rehearsal room and studio recordings, but also feature narrative elements that are not necessarily directly related to the music. Once again, it is evident that such developments would have been inconceivable without the YouTube platform, particularly given the fact that, apart from a few exceptions (Pillai 2018), music videos had only a subordinate role in the jazz sector during the MTV-era. Following the decline of music videos in the early 2000s due to the economic downturn in the music industry, which resulted in a reduction in the budget available for such productions, there has been a resurgence in recent years, particularly on YouTube (see, for example, Vernallis 2013; Dreckmann 2021; Dreckmann and Vomberg 2023; Korsgaard 2017). As evidenced by my own research (Burkhart 2025) on music videos by the band Sons of Kemet, video clips allow jazz musicians to convey sociocritical messages in an audiovisual format, circumventing the limitations of relying on lyrics or record covers, which was previously a common practice in jazz due to the prevalence of instrumental music without lyrics (Jost 1999, 61; Knauer 2020, 70). As is typical for music videos, complex intermedial reference structures are constructed that are not immediately apparent when watching the clips.

Initial research has also been conducted on jazz on TikTok. Kaye, Zeng, and Wikström (2022, 107–119; see also Kaye 2022; Kaye 2023) found that jazz-related communities have developed on the platform, with musicians – and not just fans – as the main members. The community-run TikTok channel JazzTok was launched during the global COVID-19 lockdowns, and numerous jazz-specific videos on the platform are still tagged with the hashtag #jazztok. The development of JazzTok videos was facilitated by the interactive features available to TikTok users. In particular, the *duet* feature, which allows existing videos to be duetted in split-screen mode, offered musicians the opportunity to find creative ways of interacting with other musicians in times of social isolation (Kaye, Zeng, and Wikström 2022, 114). Thus, a technical feature of the platform not only facilitated collaboration between musicians, but also contributed to the development of platform-specific music-making practices in jazz.

As Hannah Judd demonstrates, jazz-related memes have been prevalent in digital spaces for some time (Judd 2022). In her article, Judd delineates the dissemination of what was known as *the lick*. This is a brief melodic segment that has been shown to occur in numerous jazz improvisations. To identify instances of the lick, a Facebook group was established in 2010, in which group members typically shared YouTube videos of jazz recordings featuring the lick. Following the musician Alex Heitlinger's publication of a compilation of several recordings on YouTube in 2011, which included this melody segment, the lick evolved into a viral phenomenon (Judd 2022, 397). Since then, not only are jazz improvisations featuring the lick being compiled on various platforms, but numerous musicians are also playing the melody segment themselves and incorporating it into humorous contexts. For example, the musician Adam Neely published a video on his YouTube channel in which he plays the lick on an electric bass for five hours without interruption in continuous repetition (Judd 2022, 407). The lick has long been known beyond jazz circles. This example provides a compelling illustration of the potential for online media to facilitate the dissemination, creative processing, and humorous framing of jazz-specific content (for further discussion, see Galloway 2023; Spencer 2023).

Despite the dearth of literature on jazz and online media, a plethora of potential research topics has already been identified. Some of the aforementioned researchers advocate for further and more comprehensive research on jazz in digital spaces. It is posited that jazz research could benefit from this, as the observation of online media-specific jazz phenomena promises valuable insights into reception practices that would otherwise remain unexamined. As Tom Sykes argued as early as the late 2000s: "[J]azz scholars' understanding of new media and the analysis of its relationship to jazz should lead to a better understanding of jazz audiences and their modes of consumption" (Sykes 2009, 394). Ken Prouty presents a comparable perspective, contending that the examination of online discourses can advance the role of jazz studies as a means of gaining insight into contemporary debates that are not initiated or controlled by established gatekeepers:

> What virtual communities do offer is the opportunity for a wider group of jazz people to engage in this ongoing and long-standing conversation, across vast geographic and demographic spaces. For scholars of jazz, the virtual jazz world also affords us a window through which we can observe – and participate directly in – these conversations as they take place, more rapidly and more broadly than before. It's an ever-shifting work in progress, much like the music itself. (Prouty 2012a, 83–84)

While Sykes' and Prouty's arguments are compelling, this book will not focus on the sphere of jazz reception, as the two authors et al. have done in their studies. In the age of digital media platforms and the creative practices that emerge on these platforms, it seems advisable to examine the contexts of jazz-related production in digital spaces. This should reflect on the extent to which jazz is popular on specific platforms, how musicians use these platforms to present their music, and whether certain logics inherent to these platforms might even contribute to the emergence of new aesthetic practices. The research outlined above on music videos by jazz musicians, the #IHarmU videos, and the JazzTok community suggests that digital platforms can certainly influence the aesthetic practices of jazz musicians. It is imperative that we investigate the concrete effects of platform-specific socio-technical interactions between musicians and media technologies on the creative practices of jazz musicians. This investigation should include an exploration of creative processes and a description of the artistic objects that emerge in these contexts.

Previous studies on jazz and online media have lacked an in-depth reflection on the economic, technological, and political backgrounds of digital platforms. The implicit thesis of Prouty and Sykes et al. that online media would lead to an increasing democratization of jazz-related discourse must be critically re-examined in the age of digital platforms. It is important to note that the majority of the aforementioned writings were published ten to fifteen years ago, at a time when the academic discourse on the platformization of the internet was still in its infancy. Consequently, the lack of reflection on the cultural power that can be generated by platforms may, to some extent, be due to the age of the publications referenced in this chapter.

It is now widely acknowledged that platforms are commercial enterprises with their own specific functional logics, business models, visions, and worldviews. Consequently, they always operate against the backdrop of specific ideologies (Smit 2022, 478). In addition, there are the largely opaque logics of algorithmic filtering which is used by platforms to sort and prioritize content. Research in recent years has shown that these logics are anything but neutral or even inclusive, especially with regard to categories of social difference such as *race* and gender. For example, Noble (2018) and Bishop (2021) have demonstrated that algorithmic filtering is not inclusive of diverse perspectives. It is imperative to reflect on the economic, technological, and social aspects of this phenomenon in order to gain a deeper understanding of both the socio-technical transformations that occur in a platform context and the influence of platforms on music-making practices and the representation of music cultures in digital spaces.

3. Digital Platforms

3.1 What Are Digital Platforms?

The current landscape of web-based communication is dominated by a relatively small number of digital platforms, which are operated by some of the world's largest companies. The leading players in this domain are the so-called Big Five: Meta, Alphabet, Apple, Microsoft, and Amazon. The platforms operated by these companies represent key socio-technical infrastructures for consumption and communication on the internet today. They have a significant impact on the presentation and reception of information, online-based commerce, and the dissemination of cultural content (Burgess 2021, 21; Dolata 2020a, 8). This phenomenon is relatively recent, as digital platforms did not emerge in their current form until the 2000s. The rapid rise of platforms such as YouTube and Amazon began around the mid-2000s (Schrape 2021, 81). Since then, the internet has undergone significant changes. In contrast to the previous situation, where news providers, cultural creators, and even private individuals had numerous DIY online websites, content is now largely distributed in a centralized fashion across a few platforms (Burgess 2021, 21). Prior to the rise of platforms, online content was primarily produced and delivered by a relatively small number of actors and received by a much larger number of internet users. By contrast, platforms enable all registered users to create and distribute content independently, whereas the role of the online prosumer – i.e., a producer and consumer in one – was previously unheard of (Flew 2021, 50). The dissolution of strict distinctions between producers and recipients marked a significant shift in online communication, a phenomenon that had been anticipated in early research on digital platforms (e.g., Jenkins 2006; cf. Plantin et al. 2018, 296–97).

The term *platform* did not become widely accepted until the second half of the 2010s (Dolata 2020a, 9) and should still be understood as an umbrella term.

This is because a wide variety of platform types have emerged since then, with very different content focuses and business models, making it difficult to define the term in a uniform manner (Dolata and Schrape 2023, 2). Nevertheless, some overarching characteristics can be identified that characterize digital platforms on a general level. The first fundamental aspect is that platforms are digital infrastructures that operate on the basis of specific technological processes and business models, and that bring together different actors (such as companies, political actors, and private individuals) with different interests. Moreover, as van Dijck, Poell, and de Waal argue, individual platforms are always integrated into a network of multiple and interdependent platforms:

> An online "platform" is a programmable digital architecture designed to organize interactions between users – not just end users but also corporate entities and public bodies. It is geared toward the systematic collection, algorithmic processing, circulation, and monetization of user data. Single platforms cannot be seen apart from each other but evolve in the context of an online setting that is structured by its own logic. A "platform ecosystem" is an assemblage of networked platforms, governed by a particular set of mechanisms [...] that shapes everyday practices. (van Dijck, Poell, and de Waal 2018, 4)

This definition posits that the specific handling of user data – including the collection, monetization, and algorithmic processing of this data – is a defining characteristic of the business model of the companies behind the platforms. Moreover, platforms exert a profound influence on the "everyday practices" (van Dijck, Poell, and de Waal 2018, 4) of their users, as they facilitate and to some extent structure specific processes of social exchange and media consumption (Dolata and Schrape 2023, 8; Eisenegger 2021, 17). Consequently, platforms are contingent upon the content production and interaction of heterogeneous actors, including corporations and private individuals. Nevertheless, the communicative rules that prevail in the platform context are ultimately defined by profit-oriented corporations. This means that a top-down relationship between platform operators and users is fundamental to the business model of commercial platforms (Dolata 2021, 101). While certain characteristics can be identified that are constitutive of platforms in general, even a cursory examination of some of the most prominent platforms reveals their heterogeneity. Tarleton Gillespie notes that common definitions of platforms include social media sites such as Facebook and Instagram, as well as search engines (e.g., Google Search and Bing), app stores (e.g., Apple App Store and Google Play), and dating portals like Tinder or Grindr (Gillespie 2018, 254).

The various types of platforms can be broadly classified into two categories: *infrastructural platforms* and *sectoral platforms*. Most infrastructural platforms, as defined by van Dijck, Poell, and de Waal (2018), are part of the Big Five, which are the most prominent platforms in the digital landscape. The term *infrastructural* is used to describe these platforms because they form the core of the ecosystem described by van Dijck, Poell, and de Waal. Smaller platforms may only be able to function if they can access the services provided by the Big Five. The Big Five act as gatekeepers by providing *infrastructure services*, which include search engines, data servers, app stores, social networks, cloud services, and email services. For instance, numerous platform companies depend on cloud storage services such as Amazon Web Services, Google Cloud, and Apple Cloud, all of which are provided by the Big Five. Consequently, these companies rely heavily on the infrastructure services of the Big Five for their functionality. Furthermore, some of the most prominent platforms are operated by the Big Five themselves. For example, YouTube is owned by Google, while Facebook, Instagram, and WhatsApp are part of the Meta company.

Sectoral platforms (van Dijck, Poell, and de Waal 2018), on the other hand, focus on niche offerings, with individual platforms concentrating on providing news, arranging vacation rentals, or coordinating food deliveries – to name just a few examples. The reliance of such smaller platforms on the infrastructure of the Big Five stems from the integration of specific services from Google and others, but also – as mentioned above – from the use of cloud storage services. For instance, the vacation rental platform Airbnb collaborates with the online mapping service Google Maps, while the streaming services Spotify and Netflix utilize the cloud services Google Cloud and Amazon Web Services, respectively, to store vast amounts of audio and video data and make it available on demand (van Dijck, Poell, and de Waal 2018, 13). In principle, entry into the platform ecosystem is initially open to a wide variety of players. In practice, however, smaller platforms are unable to compete with the offerings of the Big Five and are generally only able to assert themselves in specialized niche markets (van Dijck, Poell, and de Waal 2018, 15).

Further distinctions can be made beyond the initial classification into infrastructural and sectoral platforms. For instance, Ulrich Dolata proposes a distinction between seven different platform types based on the primary services they offer. The first category is search platforms, where Google is the market leader, and other providers, such as Bing, are largely aligned with Google's services. The second category comprises social media and messaging platforms, with Facebook, Instagram, WhatsApp, X (formerly Twitter) and Snapchat among the most prominent examples. Thirdly, Dolata identifies media platforms such as YouTube, Netflix, Spotify, and TikTok, which are undoubtedly among the most popular platforms and are a key factor in the everyday media use of countless individuals. Fourth, some of the most widely used platforms are retail platforms such as Amazon, eBay, and Zalando. Fifth, numerous booking and service platforms have recently emerged, including in the areas of car and passenger transportation (e.g., Uber, Lyft), accommodation services (e.g., Airbnb, Booking.com, Expedia), and online dating (e.g., Tinder, Parship, Match). Sixth, there are cloud platforms such as Amazon Web Services, Google Cloud, and Apple Cloud. Seventh, there are crowdsourcing and crowdfunding platforms for companies and individuals to use to raise funds for projects, such as Kickstarter, Indiegogo, and Amazon Mechanical Turk (Dolata 2021, 101).

This overview shows that platforms have pervaded numerous domains of private and public life, in some cases assuming a regulatory function in various economic sectors and in others facilitating the development of these sectors. Consequently, the diverse activities of internet users are profoundly influenced by platforms and the companies behind them. The term *platformization* of the internet has been employed in this context for several years. This term was first coined by Anne Helmond and refers to the rise of the platform as the dominant infrastructural and economic model of the social web (Helmond 2015, 1). Alternatively, it is also used to describe the expansion of social media platforms into the rest of the web and their drive to make external web data "platform ready" (Helmond 2015, 1). The dominance of platforms described by Helmond was enabled by the establishment of direct connections – and thus the exchange of data streams – with other websites. One of the ways this was achieved was through the integration of the Facebook Like button (Burgess 2021, 22). Initially, the term platformization was understood in primarily technological terms to refer to the increasing dominance of platforms in digital spaces. However, more recent interpretations of the term have emphasized the social implications of this process. Platformization is now used to refer to the growing social significance of platforms, which is, of course, based on the technological expansion described by Helmond. Additionally, platforms are said to have a steadily increasing influence on various sectors of the economy. This is evident, for instance, in the creative and cultural sector, where creative professionals are often encouraged to be increasingly present on platforms in order to reach their audiences. Consequently, they must adapt to the regulations of the platforms. This also applies to private individuals, who frequently utilize messenger and social media platforms for a significant share of their communications (Burgess 2021, 21; Eisenegger 2021, 17). In this context, the term platformization primarily refers to the potential for platforms to exert control over users and their data, as well as the content they consume and the social interactions they engage in. This control can be exercised through various means, including controlling access to the internet, monitoring and commodifying data flows and user actions, curating content, and initiating social activities (Eisenegger 2021, 22–23).

The term *platform society* has emerged in academic discourse where it is used to describe the growing influence of platforms in shaping economic and social processes. This concept emphasizes that platforms are an integral part of modern society, influencing both economic and social processes to a significant extent. The term platform society was first introduced by van Dijck, Poell, and de Waal in 2018, who argue that "platforms are an *integral* part of society" (van Dijck, Poell, and de Waal 2018, 2; italics in original). The authors suggest that both economic and social processes are increasingly influenced by privately organized platform companies. It is crucial to differentiate between the platforms themselves, on which active participants engage online, and the companies behind these platforms. It is evident that it is not the platforms themselves that establish the rules that potentially structure action in digital spaces but that these are defined by the responsible companies before they materialize in the interfaces and algorithmic systems of the platforms and influence the actions of users (Dolata and Schrape 2023, 2).

3.2 Functional Logics of Digital Platforms

The question of how platforms build their economic and cultural power is one that requires an understanding of the specific functional logics that underpin this process. These logics can vary in detail depending on the platform in question, but there are certain mechanisms that are constitutive of the anatomy of platforms in general. Van Dijck, Poell, and de Waal state: "[A] platform is fueled by *data*, automated and organized through *algorithms* and *interfaces*, formalized through *ownership* relations driven by *business models*, and governed through *user agreements*" (van Dijck, Poell, and de Waal 2018, 9; italics in original). This summary posits that the following aspects are central to the functioning of platforms: first, platforms continuously collect data that users leave behind in the course of using certain platforms and through their online activities in general. In order to filter and sort the vast amounts of data and make them usable, platforms use algorithms to automate these complex processes. Algorithms also help platform operators learn more about specific user preferences. For instance, users of media and social media platforms such as TikTok, Instagram, YouTube, and Facebook are consistently presented with a personalized selection of content. These suggestions are generated by algorithms, and users' responses to these suggestions in turn enable platform operators to learn more about their users' individual preferences and to further refine their personalized recommendations. In addition to its technical specifications, each platform also has a specific legal and economic status. This can be either a for-profit company or, less commonly, a non-profit organization. Furthermore, each commercial platform follows a specific business model, which may be based on the sale of data collected from users to third parties or on personalized advertising offers. In addition, platform users must always agree to certain terms of use and legal agreements in order to use the platforms. Such policies define, among other things, the rights of platform companies with respect to how they handle the user data they collect. According to van Dijck, Poell, and de Waal, all these aspects must be taken into account in order to trace a specific "dynamic of platform-driven sociality" (van Dijck, Poell, and de Waal 2018, 12). In order to ascertain the extent to which certain platforms can influence users' actions, it is necessary to reconstruct the functional logics of the platforms and then relate them to the observable behaviors of their users.

In addition to the general functional logics listed above, the research literature on digital platforms also discusses other mechanisms that relate specifically to how platforms handle the data they collect, how they monetize that data, and how they select and curate content. At the core of this are the platform mechanisms of *datafication*, *commodification*, *selection/curation*, *personalization*, *reputation/trends*, *moderation*, and *terms and conditions*. Some of these mechanisms are closely intertwined and work in tandem.

3.2.1 Datafication

The term *datafication* refers to the process of extracting and collecting data about the usage behavior of human actors and social interactions in digital spaces. This data is collected by platforms, sorted, analyzed, and then used for strategic purposes (Burgess 2021, 21–22). The collection of such data is realized through both software and hardware. The devices that users use to access platforms are equipped with software and applications that enable data collection. In essence, each mouse click or cursor movement made by an internet user can be utilized to generate, store, and subsequently analyze data about the user's behavior and preferences. In certain instances, this is accomplished through the use of social buttons or pixels that are integrated into websites outside the platforms, one notable example of this being the Facebook Like button (Burgess 2021, 9). In this manner, digital platforms are now capable of transforming a multitude of areas of users' everyday lives, about which they previously lacked substantial information, into data (van Dijck, Poell, and de Waal 2018, 32). While the collection of basic socio-demographic data such as age, geographic location, and gender for market research purposes is not a novel phenomenon, the data collected by platforms is characterized by a previously unprecedented level of detail (van Dijck, Poell, and de Waal 2018, 10–11). It is noteworthy that these novel methods of data generation and analysis were initially regarded as mere byproducts of the operational business of digital platforms. However, as platform companies gradually evolved into data companies, they began to view data as a prime resource (van Dijck, Poell, and de Waal 2018, 32).

In the context of datafication, it is of paramount importance that the data trails generated by platform users in the course of their online activities are highly standardized in technological terms. This is to facilitate the automatic exploitation and use of the data across platforms (van Dijck, Poell, and de Waal 2018, 35). The data generated is not used exclusively by individual platforms; it is shared throughout a larger network consisting of media companies, advertisers, and intermediaries. Concurrently, the platforms create statistics and rankings based on the data generated, which in turn are employed to determine which content to prioritize, thereby enhancing visibility and subsequently facilitating monetization (Burgess 2021, 23). Consequently, the mechanisms of data collection and exploitation by platforms can also influence the behavior of media companies and content creators active on these platforms. These actors depend on being visible on the respective platform and responding to market and platform demands. The metrics generated by the platform can be employed by content creators to identify promising content and subsequently devise effective strategies for its success.

However, the data that can be accessed by external actors, i.e., individuals and entities that do not work for platform companies, is subject to strict control. The application programming interfaces (APIs) provided by platform companies, which allow for the controlled access to data generated in the context of the platform, play a key role here. The first API was introduced by eBay in 2000, and platforms are now equipped with APIs by default. Although content creators and media companies, for example, are provided with comparatively detailed information on the use and distribution of their content in the form of statistics, it is ultimately up to the platform companies themselves to decide to what extent outsiders are granted access to the collected data (Eisenegger 2021, 20).

3.2.2 Commodification

The business model of digital platforms is predicated on the collection and monetization of user data. By continuously monitoring user behavior, platforms gain access to detailed data that can be used to infer user preferences and behavior. However, observing a user's activities on streaming platforms such as Spotify or media platforms such as TikTok does not automatically indicate the user's current emotional state. Nevertheless, the platforms possess the capacity to observe user behavior in great detail and to establish connections between the content received and a multitude of situational factors that may influence the reception processes. Ulrich Dolata illustrates this process using the example of music streaming:

> The business of streaming services such as Spotify is based on the seamless and increasingly precise observation, evaluation and prediction of individual user behavior, which has become possible due to the enormous progress in digital surveillance technologies in the 2010s […]. Not only the search for artists or pieces of music, the playing, cancelling or skipping of songs, the creation of individual playlists and the adding or deleting of titles including date and time are automatically recorded. In addition, the platforms' collection of personal user data includes which playlists are listened to when and where, what is listened to at what times, and who exchanges information with whom. The collected and aggregated data go far beyond the creation of rather static profiles of individual users with comparatively stable characteristics (such as their basic music preferences). By increasingly including situational factors such as time of day, activity, location and environment in the data collection, it is now possible to create much more specific and context-related individual profiles that can be used, for example, to capture a user's various moods and music preferences at different times of the day or

> at different places. With all this, individual "data doubles" are created as a "reified, datafied version of the self" [...], which not only track and map the activities and preferences of platform users over time but also form the basis for predicting future user behavior. All this is far removed from classic and rough group ascriptions of musical taste along broad categories such as age, gender, class or ethnicity, which radio stations have traditionally been using to design their programs, or music companies to structure their offerings. (Dolata 2020b, 13–14)

The collected data thus provides a certain amount of insight into the everyday structures of users and is available in enormous quantities. Consequently, the business model of platforms is based, among other things, on the "*commodification of user behavior*" (Dolata and Schrape 2023, 14; italics in original). The objective is to transform online activities into tradable commodities (van Dijck, Poell, and de Waal 2018, 37). As a result of their exclusive access to this data, platform companies are continuously driving the monopolization of the data economy that they themselves have initiated (van Dijck, Poell, and de Waal 2018, 37). In the daily operations of platform companies, a process of commercial measurement and valuation of private usage patterns and social activities is underway that would not have been feasible in this form prior to the advent of digital platforms due to the lack of suitable instruments for data collection and analysis (Dolata 2019, 183).

The business model of platforms is also characterized by the fact that a heterogeneous array of actors – including end users, advertisers, and service providers – come together within the platform context, often in pursuit of commercial interests with a high degree of intensity (Helmond 2015, 2; van Dijck, Poell, and de Waal 2018, 38). Multi-sided markets have emerged on platforms as a result. Prior to the advent of digital platforms, two-sided markets were prevalent in various media industries. This phenomenon can be exemplified by the news industry, which traditionally served as a mediator between readers or viewers and advertisers, typically in the form of print advertisements or TV commercials. Consequently, large media companies have historically wielded considerable influence in the realm of advertising, as they have enjoyed exclusive access to vast audiences with the capacity to disseminate identical content and messages in a remarkably brief timeframe. In the context of digital platforms, there are still individual and very large or economically powerful companies (see the Big Five), and end users and advertisers also interact on these platforms. However, social institutions and non-corporate influencers also operate on platforms, as well as various players whose business model is based on processing the data collected by the platforms. The term *multi-sided markets* is therefore used to describe the economic interaction between heterogeneous actors in the same media environment (Nieborg and Poell 2019, 90).

3.2.3 Selection/Curation

The sheer volume of content available on various platforms makes it challenging for individual users to gain a comprehensive understanding of the resources available to them. To avoid overwhelming users and ensure a positive user experience, it is essential to pre-structure the content on these platforms in a way that presents users with a selection of content that is tailored to their individual interests and needs. Platforms are able to implement this approach on the basis of the collected data and the mechanisms for data evaluation. This results in a selection of content that is closely linked to datafication and commodification processes (van Dijck, Poell, and de Waal 2018, 40–41).

Platforms not only select content for individual users, but also help determine which topics and content are made visible to large numbers of users. This is exemplified by trends such as those that are commonplace on TikTok. Algorithmically controlled processes are partially responsible for the visibility of individuals and objects within the context of platforms. Cultural selection and hierarchization processes that were previously the exclusive domain of traditional gatekeepers such as journalists are now automated (Poell, Nieborg, and Duffy 2022, 91–92).

Nevertheless, there are still human actors involved in these selection processes, at least to some extent. This is evidenced by the fact that algorithmic content moderation, at least in the field of music streaming (Bonini and Gandini 2019), is often controlled and supplemented by human curators. Curation on music streaming platforms thus cannot be assumed to be fully automated.

The curation of content reveals a paradox in the interaction between platforms and users. On the one hand, platforms provide a vast quantity of content, as exemplified by the advent of location-independent and almost limitless access to music through music streaming services. This was touted as a unique selling point to encourage consumers to use such services. Conversely, the advent of personalized listening recommendations meant that increasingly sophisticated selection and curation mechanisms were developed over time (Dolata 2020b, 16). These recommendation lists serve as a guide to users so that they are not overwhelmed by the inexhaustible amount of choice on offer.

3.2.4 Personalization

The process of personalizing content is inextricably linked with the processes of datafication, commoditization, and, most crucially, curation as previously described. Users' reactions to the personalized content offered on platforms offer insights into the preferences and potential usage patterns of individual users, which in turn enable the collection of efficiently monetizable data. Content personalization is therefore a core mechanism, but the processes involved cannot be reconstructed in detail. This is primarily due to the fact that the functionalities of the algorithms that make these processes technically possible are trade secrets of the platform companies and are also subject to constant change, as they are continuously adapted in order to optimize functionality (van Dijck, Poell, and de Waal 2018, 41). Consequently, research on algorithmic logics is inherently constrained, as it is only possible in the most exceptional of circumstances to engage in discourse with those responsible for the technical development of platform algorithms (Bonini and Gandini 2019; Seaver 2022).

3.2.5 Reputation/Trends

Besides the personalization of content for individual users, another defining feature of platforms is the identification or creation of trends. This refers to content that spreads rapidly and often unpredictably on platforms, sometimes reaching enormous numbers of users. The factors that contribute to the emergence of trends on platforms are typically dependent on the number of users who interact with specific content within a relatively short window of time. This can include sharing, repeatedly consuming, or commenting on text, photo, or video posts. It can be argued that users exert a considerable influence on the virality of content. However, it is important to note that there is a previous step in which platform algorithms first decide which content is recommended to many users, and can subsequently evolve into a trend (van Dijck, Poell, and de Waal 2018, 41).

3.2.6 Moderation

The question of which content is allowed on platforms depends on the moderation policies of the individual platform. Moderation refers to the process of "*pre-screening, rejecting, removing, sequestering, banning, downgrading, or demonetizing content and accounts by platforms*" (Poell, Nieborg, and Duffy 2022, 96; italics in original). Platforms determine which content can be uploaded and consumed, as well as which individuals or institutions are given the opportunity to upload or be active on the respective platform. The moderation strategies employed by platforms have repeatedly led to controversy. For instance, nudity has been considered worthy of censorship on numerous occasions, while populist or racist content or content inciting violence from private individuals or political splinter groups has often escaped censorship (van Dijck, Poell, and de Waal 2018, 44–45). Similarly to curation, these processes are not fully automated. Platform companies typically employ a significant number of individuals tasked with content moderation, whose role is to remove content that has been identified as being problematic. In addition, users are sometimes encouraged to report such content (Gillespie 2018, 262–63). For instance, Douyin, the Chinese equivalent of TikTok, is said to employ approximately 10,000 moderators whose role is to monitor uploaded content and ensure its compliance with the Chinese government's censorship regulations (Poell, Nieborg, and Duffy 2022, 98).

3.2.7 Terms and Conditions

Although the precise criteria for content moderation are not publicly disclosed, sets of rules delineating what is and is not permitted on platforms are visible to the general public. Platform companies issue terms and conditions and community guidelines in which they define the social norms that all users must adhere to in order to avoid the risk of being excluded from a platform (Dolata and Schrape 2023, 8). The policies of the major platforms differ only slightly, with pornographic and violence-glorifying content, hate speech, and the depiction of drug use generally being prohibited. However, it is up to the platform companies themselves to define in detail what is to be deemed pornographic content and where the boundaries between pornography and sexualized – and therefore generally permissible – depictions lie (Gillespie 2018, 263–64).

These guidelines outline the way in which platform companies manage user data and the basic principles of user monitoring. Users must consent to these policies in order to be active on platforms. Consequently, terms and conditions are of paramount importance, as they regulate both access to platforms and users' options for action, as well as the use of their data. On platforms, users therefore always encounter *"platform-specific rules of action"* (Dolata and Schrape 2023, 12; italics in original), which, according to Dolata and Schrape, can be characterized by four key features. Firstly, it should be noted that these rules are not open to negotiation; they are defined by the platform companies, creating a top-down relationship on platforms in this respect. Secondly, the rules defined by the platform companies translate into the interfaces and algorithmic structures of the platforms and form a technical set of rules that cannot be simply overridden and always fulfil the function of structuring action. Thirdly, the rules can be continuously changed and adapted by the platforms. Furthermore, the platforms monitor all user activities within the framework of their own rules (Dolata and Schrape 2023, 12–13).

3.3 Digital Platforms and Relations of Cultural Power

The preceding analysis has shown that platforms take on the role of accentuating actors in modern media cultures in a variety of ways. They can, in principle, influence economic processes as well as processes of social exchange and cultural production in digital spaces. On the one hand, platforms create new opportunities for interaction between, for example, creative artists, advertisers, and private individuals. On the other hand, these interactions are regulated by algorithms, interfaces, moderation guidelines, and terms and conditions. As a variety of processes, including private communication, the distribution of aesthetic objects, and product advertising, increasingly take place in a platform context, individuals, creative artists, and companies are compelled to establish an online presence on platforms and to adapt to the rules of the game to some extent. In many professional fields, self-presentation on various platforms has become a basic prerequisite for economic success.

For these reasons, platform companies are initially perceived as highly influential economic actors. However, their influence extends beyond the economic realm (Dolata 2019, 183; Gillespie 2018, 254). As van Dijck, Poell, and de Waal argue: "Platforms do not reflect the social: they *produce* the social structure we live in" (van Dijck, Poell, and de Waal 2018, 2; italics in original). Even relatively early relevant researchers have argued that platforms cannot be understood in a purely technical sense, nor exclusively as digital spaces of social interaction and cultural production (Helmond 2015, 2). Rather, the specific intertwining of software-based, programmable, and algorithmically controlled infrastructures, economic commerce, and social action spaces is characteristic of the relatively new phenomenon of platforms. Dolata and Schrape state: "Internet-based platform companies represent a new form of enterprise featuring unique characteristics of *social* embeddedness" (Dolata and Schrape 2023, 2; italics in original).

Although it is now widely acknowledged that platforms cannot be viewed as neutral actors, the myth of impartial platforms has persisted for some time (Gillespie 2018, 256–57). Initially, platform companies portrayed themselves as non-interventionist, arguing that they provided access to their services to anyone with the necessary technical resources. Moreover, access to the platforms did not appear to involve any financial outlay, as no registration fees were generally charged. However, as previously described, the data trails left by users are monetized, so of course the platforms earn money with every registered and active person (Gillespie 2018, 256–57).

The concept of platform neutrality has its origins in internet-related discourses of the late 1990s and early 2000s. At the time, one of the central expectations of the so-called Web 2.0 was that it would break down the rigid roles between producers and recipients of media offerings. For example, it was predicted that journalistic mass media would lose importance as compared to user-generated content, leading to a democratization of digital spaces. In this context, considerable optimism was placed on the online encyclopedia Wikipedia, which promised to focus on collective intelligence and an emancipation from traditional gatekeepers of knowledge production and communication. In essence, Web 2.0 promised, to a certain extent, the dissolution of inequalities between large gatekeepers and media users, as well as a strengthening of democratic decision-making processes inside and outside of digital spaces (Schrape 2021, 70–72).

It is evident that the relevance of one-to-many media is declining. In principle, the number of voices that can be heard in digital spaces is greater than it was before the advent of the internet and digital platforms. However, due to the functional logic of platforms described above, it can be argued that platforms "do not merely mediate public discourse, they constitute it" (Gillespie 2018, 257). It can be argued that the platforms themselves, or rather the commercial companies responsible for them, determine which forms of interaction and participation are possible on them and which content is displayed to which users on the basis of opaque algorithmic processes. The specific forms of sociality that manifest on platforms are thus fundamentally dependent on the respective technical functional logics.

Moreover, it is widely acknowledged that the curation mechanisms of digital platforms are largely based on specific hierarchies of content. This often results in certain types of content becoming more visible than others, thereby reproducing cultural hegemonies within this context. Relevant studies have demonstrated that the results generated by search engines are sometimes based on racialized stereotypes, which are often also linked to other categories of social difference, primarily gender. For instance, it is noteworthy that certain professions that are widely regarded as prestigious, such as doctors, are predominantly associated with *white* and male individuals in the results of Google image searches. In contrast, search results for Black and female people yield highly sexualized images (Noble 2018). With regard to sexist stereotypes, for example, Sophie Bishop's study of beauty vloggers demonstrates that, in certain segments, the YouTube platform prioritizes content that is characterized by stereotypically feminine representations. Furthermore, certain keywords, such as *beauty* and *makeup*, are primarily associated with female content creators. This implies that the perpetuation of sexist stereotypes on YouTube is, at least in part, automated, and cultural biases are embedded in the platform's algorithmic architecture (Bishop 2018).

In the context of ongoing discussions surrounding the perceived impartiality of digital platforms, it is also important to acknowledge that the vast majority of prominent digital platforms, which are frequently the subject of such debates, are operated by companies headquartered in specific geographical regions. The majority of these companies are domiciled in either the United States or China, some of the sectoral platforms also in Western Europe and Russia (van Dijck, Poell, and de Waal 2018, 26; Poell, Nieborg, and Duffy 2022, 15). Platform-specific developments in the rest of the world are barely visible in the international English-language research literature, with the exception of a few instructive works on platformization in East Asian contexts beyond China (e.g., Kim and Yu 2019; Cho 2021; Park, Jo, and Kim 2023). Given that international English-language research predominantly examines platforms operated by U.S. companies, it is important to exercise caution when generalizing research findings (Nieborg, Duffy, and Poell 2020b, 5). It is reasonable to assume that certain platforms will prioritize content that is most likely to succeed in specific geographic regions with the most promising markets.

3.4 The Action-Structuring Potentialities of Digital Platforms

It is beyond dispute that platforms exert a profound influence on social processes and the representation of cultures in digital spaces. However, the heterogeneous nature of platforms precludes the formulation of generalizable statements about their social, economic, and cultural impact. Even platforms that can be categorized as belonging to a certain type, such as media, retail, or service platforms, often exhibit considerable variation at a more detailed level. The differences can be found in the interfaces and options for use, in the communication practices between users, and in the media genres that are typical for the individual platforms. The greatest reach is achieved by the large social media platforms YouTube, Instagram, TikTok, and Facebook. These platforms have a low-threshold access, and a wide variety of actors come together in these digital spaces (Dolata and Schrape 2023, 9).

In the context of digital platforms, human actors are inherently engaged with the underlying structures that shape the platforms' actions. These include predefined user interfaces and incrementally adaptable default settings, as well as various communication features (such as commenting, messaging, and similar functions). They also comprise rules about the media genres that can be created. Furthermore, the platforms impose limitations on content length, such as the maximum number of characters permitted in a text on X (280) or the duration of videos which, at least in the early days of TikTok, was restricted to 15 seconds. Users engaged in cultural work on digital platforms must learn to navigate the specific principles of these platforms if they wish to increase their visibility and success. This entails understanding how to design short texts and videos in order to reach as many people as possible or how to communicate specific kinds of content effectively. It is also important to develop an understanding of the potential functional logic of the platform algorithms based on experience and to leverage this understanding to one's advantage – with the aim of increasing visibility. Due to the specific logic of platforms, the challenges faced by content creators such as YouTubers and TikTokers are unique to each platform (Burgess 2021, 23; Duffy, Poell, and Nieborg 2019b, 2).

It is a truism of research in the sociology of technology that technologies and technical artifacts always materialize certain social norms and rules as well as the specific worldviews of the people responsible for the development of such technologies and artifacts. These norms and rules can shape or prefigure the activities of users to a certain extent (Akrich 1992; Dolata 2013, 32–34).

It is evident that platforms can also exert a structuring influence on the actions of cultural workers. A number of media genres that have emerged in the context of platforms should be considered here, for example Instagram stories. Since this is essentially an audiovisual media genre, Instagram stories cannot be described as a new phenomenon per se. Rather, it is the specific possibilities for embellishing Instagram stories that are to a certain extent pre-figured by the specifications of the Instagram platform and that ensure recognition value or encourage users to follow certain design conventions when producing stories. These pre-configured elements include, for example, the rule that videos can only be recorded in portrait format. Furthermore, Instagram stories are distinguished by their time constraints, with content creators utilizing them to engage with followers in a distinctive, often more intimate manner than is possible with standard image or video posts.

Nevertheless, it is irrefutable that such structures and rules inscribed in technologies are never determinative of action; they can always be interpreted to some extent (Dolata 2019, 198). To illustrate: let us consider, again, the example of Instagram stories. Within a clearly defined framework, the specific design of these stories is always the responsibility of the human actors on the platform. This is an essential prerequisite for the emergence, establishment, and further development of new media genres. This would be implausible if users were not afforded the opportunity for creative interpretation of the given media genres. In this respect, platforms are also dependent on users, who develop new practices in the context of the platform and establish communicative norms that cannot be fully foreseen by the platform operators. Therefore, a recursive relationship between platforms and their users must always be assumed. On the one hand, platforms influence the behavior of users; on the other hand, users also influence the development of platforms (Duffy, Poell, and Nieborg 2019b, 2; van Dijck, Poell, and de Waal 2018, 11).

3.5 Analyzing Socio-Technical Relations in the Platform Context

The preceding sections have made it evident that the specific relationships between platforms and human actors are a central topic of platform-related research. In essence, these are *socio-technical* interactions based on the interplay between platform mechanisms and user practices. Practices of human actors that are linked to specific software and hardware configurations can become highly normalized or habitualized over time. One illustrative example is the use of a keyboard or a touch screen, where the seamless functioning of the devices and the proficient handling of these technical artifacts by the users must be guaranteed (Schrape 2021, 33). This results in the formation of routines that would not be feasible without the intertwined involvement of human and non-human actors and that are equally contingent on both the technologies and their human interpretations.

Such socio-technical relations are also crucial in the digital spaces that emerge in the platform context and in which users interact with the platforms' algorithms and interfaces. In recent research literature, several theoretical concepts have been developed that can be used to conceptualize specific socio-technical relations in the platform context and provide an adequate terminology to describe them. The following concepts are particularly relevant for the present study: *algorithmic culture*, *affordances* of digital platforms, *algorithmic imaginaries* of platform users, and *platform vernaculars*. In each case, albeit from disparate vantage points, the inquiry concerns the manner in which human and non-human actors interact in digital spaces and the cultural practices and aesthetic objects that emerge from the socio-technical interaction processes between platforms and users.

3.5.1 Algorithmic Culture

It is beyond dispute that algorithms play a pivotal role in digital platforms. They facilitate processes such as the collection and analysis of vast amounts of data and the automated sorting, hierarchization, and personalization of content. People interact with algorithms on platforms in various ways. For instance, platform users receive algorithmically generated and personalized content recommendations. They can attempt to influence the future generation of automated suggestions by interacting with these recommendations in a targeted manner (see above). Professional content creators, on the other hand, can attempt to comprehend the algorithmic logic of various platforms to the greatest extent possible. This will assist them in developing content that enhances their prospects of maintaining visibility in the digital domain. The utilization of specific hashtags or keywords may also be beneficial in helping one's own content stand out from the crowd or disseminating it to a vast audience with specific interests that can be linked to those hashtags and keywords.

The question of what algorithms are and what they do is the subject of sometimes differing opinions in different scientific fields. In the natural sciences and informatics, algorithms are understood in a technical sense and described as automated instructions that guide computers in the execution of certain tasks. In the humanities, however, the focus is not on technical details, but on the extent to which algorithms can influence social or cultural processes. *Algorithmic cultures* are distinguished by the intimate intertwining of technical and cultural dimensions. Striphas's definition, which emphasizes this interconnection, is as follows:

> [A]lgorithms are best conceived as 'socio-technical assemblages' joining together the human and the nonhuman, the cultural and the computational. Having said that, a key stake in algorithmic culture is the *automation* of cultural decision-making processes, taking the latter significantly out of people's hands. (Striphas 2015, 408; italics in original)

In order to gain a fuller understanding of the role of algorithms in social and cultural contexts, it is essential to recognize their intrinsic relationship with digitalized media cultures. This relationship is not merely one-sided; rather, it is a two-way street, with human practices influencing the development of algorithms and algorithms influencing human practices. This reciprocal relationship is particularly evident in the context of platform development, where algorithm developers must observe how humans interact with algorithms in order to optimize their functionality. This is why algorithms, especially in the context of digital platforms, should not be understood as purely technical entities isolated from social contexts. Rather, they should be regarded as a constitutive component of social, digitized worlds (Beer 2017, 4). Roberge and Seyfert state:

> Algorithms have expanded and woven their logic into the very fabric of all social processes, interactions and experiences that increasingly hinge on computation to unfold; they now populate our everyday life, from the sorting of information in search engines and news feeds, to the prediction of personal preferences and desires for online retailers, to the encryption of personal information in credit cards, and the calculation of the shortest paths in our navigational devices. (Roberge and Seyfert 2018, 1)

Algorithms play a pivotal role in digital spaces, contributing to the sorting and hierarchization of content. Historically, making certain information, cultural objects, or groups of people visible (or invisible) in media contexts was the responsibility of human actors, often those in powerful positions within the media industry. While these traditional gatekeepers have not suddenly become obsolete due to the increasing algorithmization of social worlds, there have been significant shifts in cultural power relations (Striphas 2015, 396; Beer 2013, 97). It is important to note that algorithmically driven sorting and hierarchization processes do not take place independently of the worldviews of the people responsible for developing the algorithms. The development of algorithms is inevitably influenced by certain cultural and social patterns that are inherent to the technical systems created by human actors. For instance, the notion that certain groups of users are more likely to engage with certain content on specific platforms can influence the functional logic of algorithms (Gillespie 2014, 177; Cotter 2019, 898). In this context, the perpetuation of cultural hegemonies through algorithms is inevitable. Sophie Bishop posits that the field of developing algorithmic systems is still male-dominated, which means that gendered stereotypes automatically influence the way these systems function (Bishop 2018, 71). Although the exact details of these processes are challenging to reconstruct, it seems at least evident that the hierarchization of algorithmically moderated spaces is inextricably linked to the influence of human actors (Slack and Hristova 2020, 22).

The concept of algorithmic culture refers to the intricate intertwining of human and non-human actors in digital spaces. According to Slack and Hristova, "foregrounding algorithmic culture requires addressing the *connections* that constitute what matters most about algorithms: their integration into practices, policies, politics, economics, and everyday life, with consequential political, ethical, and affective meanings" (Slack and Hristova 2020, 16; italics in original). From this perspective, then, the key question is the extent to which algorithms become part of social realities, influencing human action but not predetermining it.

3.5.2 Platform Affordances

Platforms create digital and algorithmically moderated spaces in which algorithmic cultures emerge. In this environment, a variety of sociotechnical interaction processes occurs between human actors, the technologies underlying the platforms, and the artifacts required to use the platforms (e.g., smartphones, tablets, notebooks, etc.). Users are presented with a range of content that has been algorithmically moderated. The interfaces of the platforms and the respective media genres (e.g., short-form videos on TikTok, short texts on

X) offer certain courses of action which are in turn interpreted by human actors and can form the basis for creative practices. The concept of *platform affordances* explores the action-structuring potential of platforms, which can guide the behavior of users to a certain extent. In other words, the focus is on the *in-between* between platforms and users.

The term affordance was first coined by psychologist James J. Gibson, who originally developed the concept as part of his research on the relationships between non-human animals and the environment (Gibson 1979). In essence, the term refers to the options for action that a particular environment offers to a particular subject, although it is important to note that these options are not fixed but rather vary depending on the individual subject's abilities. For instance, a tree offers disparate options for different non-human animals to seek shelter, climb, or find food, despite the tree's physical properties remaining constant. Affordance is thus always to be understood as a relational category. The options for action that an object offers to an individual are neither entirely predetermined by physical properties, nor entirely socially constructed or arbitrary.

In the following decades, building upon the work of Donald A. Norman (1988) in the field of design research, the concept was further developed – especially in research on material culture. The central question is which human behaviors are more likely to occur when interacting with specific objects, and to what extent this is related to the material properties, design, and technical functionality of the objects themselves. The concept of affordance has become a central tenet in a diverse array of disciplines, including media sociology (Zillien 2008) and archaeology (Fox, Panagiotopoulos, and Tsouparopoulou 2018). It is also a key focus of research examining the agency offered to users by technical music devices like record players (Hoklas and Lepa 2015) and digital audio workstations (Bell 2015).

The term has been the subject of discussion in the field of digital platform research for some time, and it is conceptualized in a relatively broad sense. For example, affordance does not necessarily refer to the function of a single button (such as the Facebook Like button). Rather, the question is what communicative practices are enabled, challenged, or suppressed by such a button. In this respect, the term in the context of platforms refers to the "'multifaceted relational structure' […] between an object/technology and the user that enables or constrains potential behavioral outcomes in a particular context" (Evans et al. 2017, 36). In particular, it prompts the question of how certain platforms facilitate certain user behaviors and thus the production of certain content on the platforms, most likely due to specific technical and communicative features (Bucher and Helmond 2018, 235). Affordance should therefore also be understood as a relational category in this area of research. While the TikTok platform, for example, offers or prevents specific options for action due to its restriction of media formats to short-form videos (only videos in portrait format with a certain duration can be uploaded), the *actual* modes of use only emerge in the context of active appropriation by users and can therefore be quite heterogeneous (Hopkins 2020, 48; Ilten 2015, 5).

With regard to the presentation of music cultures on certain platforms, it is necessary to consider which musical repertoires can be adequately presented in the context of platform-specific media genres. In order to determine how musicians engage with platform-specific affordances, it is essential to understand which facets of music cultures are more likely to succeed on platforms than others. It is reasonable to posit that as people interact with platforms, certain strategies employed by creative artists will evolve in response to the platforms' functionalities and the opportunities they afford. These strategies may be employed with the intention of enhancing visibility.

3.5.3 Algorithmic Imaginaries

A competition for visibility is taking place among content creators on digital platforms. This phenomenon has recently been described as a "popularity contest" (Bucher 2018, 105), for example, or a "visibility game" (Cotter 2019, 896). Users who engage in the production of content and who wish to differentiate themselves from their peers and achieve visibility must identify strategies for leveraging the algorithmic systems of digital platforms to their advantage. In some instances, this may necessitate adapting their content to align with the logic of the platform, with the goal of reaching as large an audience as possible. It is evident that generating media visibility has been of significant importance to cultural workers for a considerable amount of time prior to the advent of digital platforms. Indeed, mass media have long held a powerful position as communicators and gatekeepers (Bucher 2012, 1165). However, the framework has undergone a significant transformation. Rather than focusing solely on human actors, media content must now also address the algorithmic systems of the platforms (Bucher 2018, 111). Consequently, cultural workers now face the challenge of making their content "algorithmically recognizable" (Gillespie 2017, 65) on these platforms.

In order to implement such optimization processes, individuals must develop a specific understanding of the logic underlying the platforms' algorithmic systems. They then use this understanding to develop specific strategies for action and to tailor their content accordingly (Gillespie 2014, 184). However, the precise functional logics of algorithmic systems remain opaque to individuals not employed by the platform, including successful content creators, as these logics are trade secrets of the platform companies. The resulting strategies are thus inherently conjectural (Bishop 2019, 2591). Hence, it is not only the platform algorithms, which are technical systems, that occupy a powerful position in the competition for visibility; the content creators' ideas about what algorithms *could* do also develop their own agency, which should not be underestimated (Beer 2017, 11). The manner in which individuals interpret the functional logic of algorithms can influence their behavior on platforms in ways that are difficult to predict or preclude in detail by platform companies (Cotter 2019, 896).

Taina Bucher has proposed that human ideas about how algorithms function can be understood as *algorithmic imaginaries*. According to Bucher, such imaginaries about algorithms should be understood as "ways of thinking about what algorithms are, what they should be, how they function, and what these imaginaries in turn make possible" (Bucher 2018, 113). According to Bucher, content creators also need to develop a certain gut feeling for how algorithms work:

> The practical engagement with social media platforms as lived-environments implies developing tacit knowledge about the underlying logic of the system. While most technologies are designed in such a way that people do not have to know exactly how it works [...], people tend to construct "mental models" and theories about its workings as a way of navigating and interacting with the world. (Bucher 2018, 114–15)

Although the algorithmic imaginaries of individual users may differ in detail, Bucher's research indicates that content creators can reach comparable conclusions about the functioning of algorithms by specifically observing the popularization processes on individual platforms. This can result in a homogenization of content on platforms if a large number of content creators develops similar ideas about what content can be used to generate visibility in digital spaces and the production logics of different actors are adapted accordingly (Bucher 2018, 105–06). Specific strategies and, as a result, specific aesthetic objects and communicative practices can emerge on individual platforms over time which are distinctive to the platforms on which they evolved.

3.5.4 Platform Vernaculars

In the mid-2000s, Jean Burgess coined the term *vernacular creativity* (Burgess 2006). This term refers to the phenomenon whereby new communicative practices emerge in response to the influence of new media and then merge with familiar forms of communication in unique ways. For instance, within the specific media environments that platforms such as TikTok, Instagram, and YouTube offer users for the production of (audio-)visual content, certain practices emerge that refer to long-established media formats (photographs and audiovisual media), yet which can only be found in their specific form on the respective platforms. The concept of *platform vernaculars* addresses the development of such platform-specific conventions (Gibbs et al. 2015; Eriksson Krutrök 2021). Gibbs et al. define platform vernaculars as follows:

> Platform vernaculars are shared (but not static) conventions and grammars of communication, which emerge from the ongoing interactions between platforms and users. While platform vernaculars are particular to social media platforms, it is also important to acknowledge that they can share many elements, and the vocabulary and grammars of vernaculars migrate between social media platforms as new practices and features from one platform are appropriated for use on others. [...] Platform vernacular draws attention to how particulars [sic!] genres and stylistic conventions emerge *within* social networks and how – through the context and process of reading – registers of meaning and affect are produced. This approach allows us to examine the specificities of social media platforms. (Gibbs et al. 2015, 257–58; italics in original)

These vernaculars are thus created on the basis of interactions between platform users and the affordances of the platforms. Scolere, Pruchniewska, and Duffy demonstrate through interviews with content creators that decisions about which types of content are shared on which platforms depend on both the technical features of the platforms and on specific assumptions made by content creators about the cultural characteristics of different platforms (Scolere, Pruchniewska, and Duffy 2018). While different media platforms, such as TikTok, Instagram, and YouTube, function in a fundamentally comparable manner, they are also distinguished by their unique interfaces, media genres, and possibilities for action. Furthermore, they are utilized by different user groups with varying intensities. The development of specific platform cultures is influenced by a complex interplay of these heterogeneous factors (Burgess 2021, 25). Consequently, the homogenization of content on certain platforms is highly probable, as successful content creators align themselves with already established platform-specific conventions. Furthermore, due to their high visibility on platforms, they can contribute to the consolidation of such regularities in the creation of content for specific platforms or the expectations of numerous other users.

3.6 Platforms and Cultural Production

It is evident from the paragraphs above that platform-specific modes of expression and conventions of representation can emerge as a result of socio-technical interactions between platforms and users or content creators. For several years, both journalistic and academic discourses have asserted that digital platforms exert a profound influence on the processes of cultural production. In the field of music, this is particularly evident in discussions about how the functional logics of streaming platforms influence the processes of professional music production. First and foremost, there are concerns that the streaming economy will have a significant homogenizing effect on music production. This is based on the assumption that music creators will have to adapt their productions to align with the business models of the platforms. One particularly popular hypothesis is that songs would have to capture listeners' attention immediately, for instance by employing a catchy chorus at the beginning, due to the distribution mechanisms of the streaming platform Spotify, which only remunerates the creator for a stream after a listening duration of thirty seconds. Furthermore, the length of songs would gradually diminish as listeners listened to more songs in less time. Given that each stream is remunerated individually, it is assumed that greater profits could be generated in this way. However, such hypotheses are usually not based on empirical findings. Instead, they are comparatively generalized statements by single authors, often with a clear critical connotation (see Hesmondhalgh 2022 for a critique of these debates).

When viewed through the lens of sociology of technology, the skepticism about the supposed influence of digital platforms on cultural production processes is not surprising. The phenomenon of platformization, and more generally digitalization, is relatively recent, yet it has already permeated numerous sectors of society, as previously outlined. The establishment of such fundamental technologies is contingent upon heterogeneous and often protracted processes of social appropriation and adaptation. These processes are accompanied by debates about the potential socio-economic and socio-cultural consequences of these processes of change. While a long-established fundamental technology such as electricity is now so deeply embedded in social practices that it may seem *quasi-natural* to most people, the various information and communication technologies that are crucial to platformization are not yet embedded in everyday lifeworlds in a comparable way (Schrape 2021, 10–11). For these reasons, some of the developments connected to the increasing platformization of society are currently still associated with uncertainty and fear.

However, there are other discussions of the potential influence of digital platforms on the processes of cultural production that lack this critical undercurrent. The focus here is on the extent to which cultural workers must tailor their products to specific platforms in order to comply with the respective logics of popularization, i.e., to generate visibility, clicks, likes, etc. It is argued that the rules and principles that are defined by platform companies and materialize in the platform interfaces and algorithms influence the work of content creators in specific ways. In this context, researchers often speak of *optimizing* aesthetic objects such as songs, images, or texts for specific platform contexts (Morris 2020; Morris, Prey, and Nieborg 2021; Raffa and Pronzato 2021). In the sense used by Morris, Prey, and Nieborg, the term *optimization* does not necessarily imply an increase in artistic quality but refers instead to the fact that, in the context of the platform, cultural creators must differentiate themselves from the vast quantity of content and creators in order to garner attention. In essence, aesthetic objects must be made "more searchable, discoverable, usable, and valuable in both economic and cultural senses" (Morris, Prey, and Nieborg 2021, 162–63). In this respect, it would make sense to consider pictures or songs, for example, as datafied objects to a certain extent. Everyone involved in the production and distribution of these objects (in the music sector, for example, musicians, label employees, producers) would have to some extent become data scientists in order to succeed in the digital competition for visibility in the long term (Morris, Prey, and Nieborg 2021, 163). Although not all of the above concepts for analyzing socio-technical relations in the context of platforms are mentioned in the corresponding writings, it is evident that the fundamental ideas regarding algorithmic cultures and imaginaries as well as platform affordances and vernaculars, are reflected here.

In the context of concrete musical optimization processes, reference is sometimes made to the production of "streaming-friendly music" (Nordgård 2021, 46), which is music that is assumed to be more successful on certain platforms due to its being tailored to specific algorithmic filtering processes. Such theories often refer to music that is distributed on particularly popular Spotify playlists and is designed to make no demands on the listener's attention, so that it can play more or less unnoticed in the background. This would allow multiple songs to play one after the other and be paid for accordingly (Nordgård 2021, 46). With regard to the theory that songs tend to become shorter due to the distribution logic of Spotify and comparable streaming platforms, Morris (2020) even assumes specific *platform effects* in the field of contemporary music production. This term is a reference to the *phonograph effect* described by Mark Katz (Katz 2010), which Katz attributes to the development of sound recording technologies in the early twentieth century: musicians had to adapt the duration of their recorded music to a maximum possible playing time of approximately three minutes, which was the uppermost limit of what could be stored on one side of the shellac records that were widely used at the time. Furthermore, it can be postulated that the music recorded on sound carriers probably differed in numerous ways from the real sound events experienced at the time. It is evident that the recording methods of the time forced musicians to stand in a group in front of a single recording funnel and attempt to record all their instruments at an appropriate volume. Given that the volume ratios between the instruments were sometimes adjusted in order to enhance the quality of the recording, it is logical to conclude that this would inevitably result in discrepancies in the sound when the same music was performed live. Such observations lead to the conclusion that sound recording and playback technologies have had a significant impact on the sonic design of recorded music for some time.

The platform effects assumed by Morris refer not only to all sonic transformation processes, but also to the platform-specific optimization of musical products in a broader sense. Thus, not only does "sonic optimization" (Morris 2020, 5) play a role, but also "metadata optimization" (Morris, Prey, and Nieborg 2021, 164), which posits that musicians, label managers, and music producers must also consider using specific keywords, such as the names of well-known musicians, in song titles to generate interest and attract the platform algorithms. This assumption also pertains to typical search engine optimization strategies, such as clickbait tactics, which attempt to persuade as many users as possible to click on videos, articles, or images by using sensational headlines or attention-grabbing keywords (Morris, Prey, and Nieborg 2021, 164).

While these considerations appear to be essentially plausible at first glance, there is not enough solid empirical data to substantiate all of these hypotheses. For instance, Morris' statements, which appear to be a factual report of the strategies of professional musicians operating on platforms, ultimately concede that empirical studies will be necessary in the future to show the extent to which musicians, producers, or label managers actually respond to the supposed optimization constraints (Morris 2020, 8). There exists a problematic tendency among academic publications to uncritically adopt certain theories on music-related optimization which are regularly disseminated in journalistic texts, usually without any empirical basis. This is exemplified by an article by Diana Zulli and David James Zulli on mimetic practices on TikTok. In their article, the authors posit that the production of popular songs is subject to substantial processes of change due to the logics of the TikTok platform, particularly its focus on very short videos of fifteen to sixty seconds. They argue that it is now crucial to produce songs that are suitable for dance challenges. Zulli and Zulli conclude: "Bridges and choruses of popular music are being shortened to fit TikTok's fifteen- to sixty-second video limits. Lyrics and tunes are being designed with corresponding movements with hopes that TikTok users will attach a song to a dance challenge" (Zulli and Zulli 2022, 1884). These assumptions may appear plausible; it is, after all, reasonable to assume that certain platform effects will occur over time, or have already occurred, as creative artists must respond to changing media and economic conditions. However, such theories lack any empirical foundation, which is why the approach of formulating supposedly universally valid facts about platform-related production processes is particularly misleading in this context (see Raffa and Pronzato 2021, 301–02 for a comparable approach). Furthermore, the authors do not specify which forms of popular music they mean. This suggests that popular music cultures in general are affected by the processes of change posited by Zulli and Zulli. However, it seems more probable that this is only relevant in very specific segments of the field of what is generally referred to as mainstream pop.

In exceptional cases, music production professionals engage in research studies to provide empirical evidence to support the hypotheses of potential platform effects. As Morris, Nieborg, and Prey posit:

> Cultural producers now make cultural goods explicitly with search engines, platform economics, and discovery algorithms in mind – or at least with the

> *perception* of these regimes, models, and frameworks. [...] Given that the display, search, discovery, and consumption of cultural goods now all take place through the same software platforms that distribute cultural content, content needs to be crafted with the mechanics and infrastructure of the platform in mind. (Morris, Prey, and Nieborg 2021, 163; italics in original)

Once more, the impression is given that these statements are based on well-founded knowledge or even an insider's perspective on platform-related production processes. At the same time, the authors point out the difficulty of proving an actual causal connection between certain platform logics (especially those of the streaming platform Spotify) and processes of change in the field of music production. Consequently, interviews were conducted with "musicians and music industry insiders" (Morris, Prey, and Nieborg 2021, 164) and it was found that strategies related to "'songs that stream'" (Morris, Prey, and Nieborg 2021, 164) are frequently discussed. To substantiate this, a Dutch musician and producer is quoted anonymously:

> I know for a fact that in the New Music Friday playlist (on Spotify), a user, I think, listens to a song for about 5 seconds before they skip. So you have to catch their attention in 5 seconds. So whenever I do a session with musicians, I try to make the intro as fast or as interesting as possible. So from the top it should grab your attention. (quoted in Morris, Prey, and Nieborg 2021, 164)

This statement serves to reinforce the authors' hypothesis, although no other music practitioners are referenced in the course of the article, which is disappointing given the reference to several interviews with "musicians and music industry insiders" (Morris, Prey, and Nieborg 2021, 164). This shows that even research that is purportedly based on empirical data is unable to substantiate the hypotheses on potential platform effects.

Such observations are indicative of a significant limitation in the current diagnostic framework for digitization and platformization. While it is evident that platformization has a profound impact on social structures, the concrete effects of platformization on cultural life remain poorly understood due to the ongoing nature of the process. Such diagnoses are frequently based on subjective experiences or observations of highly specific dynamics in narrowly defined fields. Consequently, they often cannot be generalized (Schrape 2021, 49).

There is a growing number of critical voices on the alleged effects of platformization on musical production and reception. These voices are calling for an empirical basis to support such theories, which are informed by a diagnosis of the times. According to a recurring argument, it is crucial to reconstruct the actual ways in which human actors engage with digital platforms in terms of music production and reception. It is important to recognize that users do not necessarily utilize platforms in the same manner as platform companies may have initially envisioned during the development process (Hesmondhalgh 2022, 15; Jansson 2023, 3209). Consequently, while platform-deterministic approaches should be avoided, researchers must also prioritize investigating specific cultural domains within the platform context (Poell, Nieborg, and Duffy 2022, 19–20). This is necessary in order to substantiate previous theories, which are prone to over-generalization when it comes to the supposed effects of platformization on music creators. In order to do so, it is necessary to present detailed findings on how musicians in specific fields on different platforms are actually influenced by specific platform logics (Poell, Nieborg, and Duffy 2022, 4). In the words of Nieborg and Poell: "The challenge ahead is to develop in-depth case studies of how platformization unfolds in particular geographies, fields, and instances of cultural production" (Nieborg and Poell 2018, 4288).

3.7 Empirical Research on Cultural Production on Platforms

What are the primary motivations for cultural workers to become active on specific platforms with specific content? What strategies do they develop to align with the underlying logic of these platforms? And how do such strategies manifest in the aesthetic objects that circulate on different platforms? These are the essential questions that must be addressed in (not only) music-related research on platforms if we wish to generate substantial insights into the socio-technical relationships between cultural workers and platforms.

The processes of cultural and media production have been extensively researched for some time, particularly in the fields of production studies and media industry studies. The analysis of working contexts in specific media industries and the corresponding processes of cultural production can now look back on a history of approximately eighty years (cf. Caldwell 2013; Vonderau 2013; Vonderau 2023). The key question in this field of research is why certain cultural objects emerged in certain historical phases and under certain economic, technological, social, and cultural conditions, subsequently becoming popular, while others remained virtually unknown or did not emerge at all. A significant body of relevant research has focused on the impact of television and film (e.g., Banks, Conor, and Mayer 2016; Caldwell 2008; Mayer, Banks, and Caldwell 2009). Recently, this research has expanded to encompass the effects of platformization (Sundet 2021). In the context of music-related research on production cultures, researchers have thus far focused primarily on work processes in the recording studio and the network-like connections between people involved in the artistic, technical, and economic aspects of creating music recordings. These areas of research are described with terms such as "The Art of Record Production" and "The Musicology of Record Production" (e.g., Frith and Zagorski-Thomas 2012; Zagorski-Thomas 2014; Zagorski-Thomas et al. 2019). Researchers in this field typically employ established qualitative research methods, for instance by accompanying, observing, or interviewing subjects engaged in specific music-related production processes during their day-to-day work. These approaches are employed to empirically reconstruct the specific practices of those responsible for the products, thereby mapping the conditions under which aesthetic objects are created.

In principle, such approaches can be beneficial for an investigation of production cultures within the context of digital platforms. However, it is essential to first determine which groups of individuals to consider as potential interview partners in such research projects and whether researchers can effectively engage with these individuals to gain access to their working contexts and knowledge. Given that aesthetic objects in the platform context are always created on the basis of socio-technical relations between platforms and cultural practitioners and are sorted and curated by platforms, a dialogue with individuals responsible for the platforms would be a valuable means of gaining insight into the background of specific production cultures influenced by certain platform logics. This could include executives of platform companies as well as software developers and engineers responsible for the development of platform algorithms, or content curators and moderators. Nevertheless, contacting these individuals is often fraught with difficulties and is rarely feasible. This issue has been extensively discussed by Bonini and Gandini in their article on music curation on Spotify (Bonini and Gandini 2019). The authors employed a classical production studies approach in order to reconstruct the logics of music curation, which involved contacting, interviewing, and shadowing Spotify employees in their daily work. However, it became evident that Spotify as a company did not permit such research. Despite repeated attempts over several months, it was not possible to gain insight into the work of Spotify curators. The difficulty of gaining access to certain areas of the media industry has long been a fundamental problem of production-based research, which, according to Bonini and Gandini, is further compounded in the case of platform companies (Bonini and Gandini 2019, 2). It is also pertinent to mention the research conducted by Nick Seaver, who was able to gain insights into the work of the individuals responsible for the development of Spotify's music recommendation algorithms (Seaver 2022). It is important to note that Seaver's data was collected between 2010 and 2013, a period before Spotify's transition from a music database to a provider of a "branded musical experience" (Prey 2020, 2). The applicability of his findings to the current situation is therefore limited. Given the current circumstances, it is unlikely that researchers will be able to gain similar access to this field of work. Bonini and Gandini therefore set out to identify individuals who had previously worked in the field of music curation but were no longer employed by platform companies. Additionally, interviews were conducted with individuals occupying various roles within the music industry, including those engaged in music management and software development. These individuals were also asked for their perspectives on music curation (Bonini and Gandini 2019, 4). According to Bonini and Gandini, this detour is crucial for gaining insights into platform-related work processes without being dependent on the willingness of platform companies to provide information (Bonini and Gandini 2019, 8).

This indicates that platform-related production research cannot be conducted by contacting leading employees, programmers, or content moderators of the platforms. It is more useful to seek dialogue with individuals who possess specific knowledge of the functional logic of particular platforms, but who do not work for the companies in question. As in the case of Bonini and Gandini, these individuals might be former employees of platform companies. Contacting successful content creators might also be plausibly useful. We can assume that such individuals have worked extensively with specific platform logics over an extended period of time and have developed strategies for becoming visible and successful on these platforms. As a result, they are highly familiar with certain functional logics of the platforms, but do not work for the platform companies and are therefore not subject to any confidentiality obligations. It is important to note that even successful content creators may not possess detailed knowledge of the functional logic of the platforms' algorithmic systems or platform-specific curation processes. However, based on their experience, they are able to provide insight into the strategies for success they have developed in the context of the platform. This promises valuable insights into how human actors *actually* interact with platforms.

Bonini and Gandini, as well as Seaver, employ established methodologies. Qualitative research methods such as expert interviews and participant observation are well-suited for reconstructing the knowledge of individuals engaged in platform-related production. Indeed, this field of research necessitates such approaches, as specific insider information can only be obtained through direct dialogue or by shadowing cultural workers (Nieborg, Duffy, and Poell 2020, 2–3).

Empirical case studies in the field of platform-related production research are currently scarce, although some instructive work has been published in recent years. In particular, two special issues of the journal *Social Media + Society* (Duffy, Poell, and Nieborg 2019a; Nieborg, Duffy, and Poell 2020a) should be mentioned here. The collected studies are characterized by their heterogeneity in terms of content and their methodological homogeneity. The cases examined include live streamers on Twitch (Johnson and Woodcock 2019), BookTubers (Tomasena 2019), webtoon producers (Kim and Yu 2019), and LGBTQ influencers (Duguay 2019). All of these studies employed qualitative methods, primarily interviews. This suggests that heterogeneous production fields have emerged on different platforms, but that these can generally be adequately explored using qualitative methods.

For instance, in their study of Chilean fashion and lifestyle content creators, Arriagada and Ibañez demonstrate that production strategies in this area are highly contingent upon the logics of the Instagram platform and the creators' respective interpretations of these logics. The interviewees adapt to the platform-specific styles of communication, for instance, taking great care to provide their followers with regular glimpses into supposedly private contexts. Moreover, advertising revenue is always measured in terms of reach, which underscores the necessity of developing specific strategies for engaging with the platform's algorithmic systems, with the objective of making their content more visible (Arriagada and Ibañez 2020, 9). In a study of LGBTQ influencers on Instagram, Duguay posits that a relatively homogeneous image of this group of people has emerged on the platform. As the author demonstrates through interviews with content creators, this is primarily due to the fact that the product managers have engaged extensively with the logic of the platform and subsequently developed promising strategies for success on the platform. This entails not only reproducing certain staging practices that have proven successful, but also utilizing popular hashtags. Content creators are largely guided by the platform-specific challenges that arise from the combination of Instagram's guidelines and user reactions (Duguay 2018, 106).

The findings of these two studies, briefly outlined here, demonstrate that specific socio-technical relations can emerge on digital platforms in certain niche segments. As is common in algorithmic cultures, human actors are to some extent dependent on the platform's algorithms, and they also have to engage with the media genres provided by the platforms and the aesthetics of self-presentation conventionalized on certain platforms. Algorithmic imaginaries must be developed in order to increase visibility and learn how to cope with the platform's affordances. As a result, platform-specific and niche-dependent presentation conventions emerge, which ambitious content creators must adhere to in order to be successful in the long run. These processes can only be reconstructed in detail in close dialogue with content creators. Few such studies have been carried out in the field of music research to date, but there are signs of such research approaches emerging. This is necessary in order to empirically validate hypotheses about supposed platform effects and streaming-friendly music. In doing so, it is crucial to reflect on the functioning of those platforms whose influences on processes of cultural production are to be examined. In the context of the present study, it is first necessary to elucidate in detail the specific platform logics that characterize TikTok. This will inform the development of a concrete research design, which will then be employed to empirically analyze the representation of jazz on the platform, investigating the contexts of production of cultural workers in the field of jazz on TikTok.

4. TikTok

4.1 On the History of Short-Form Video Platforms

TikTok is a digital platform that enables users to upload and share short-form videos. Similar platforms, which to some extent facilitated the advent of TikTok's considerable success, have been in existence since the first half of the 2010s. While each platform exhibits distinctive characteristics, there are a number of common features that are inherent to short-form video platforms in general. It is evident that the format of the short-form video is a central aspect of all platforms of this nature, although the specifications regarding the duration of the videos vary considerably. For instance, the discontinued platform Vine hosted videos with a maximum duration of six seconds (Kaye, Zeng, and Wikström 2022, 20). Another common characteristic of short-form video platforms is the "endless scroll feature" (Kaye, Zeng, and Wikström 2022, 21). This feature means that each user is presented with a bespoke selection of videos, algorithmically curated by the platform. Should the user find the video unappealing, they can simply swipe it away, prompting the immediate display of another video. On the platform TikTok, this is referred to as the *For You page*. This feature is typically the primary means by which users discover new content on short-form video platforms (Kaye, Zeng, and Wikström 2022, 21). Furthermore, short-form video platforms are typically low-threshold, facilitating the creation, editing, and posting of videos for users with varying levels of expertise. These processes are integrated into video production features, whereas YouTube videos are produced and then uploaded without the assistance of features offered by the relevant platform (Kaye, Zeng, and Wikström 2022, 22). This option is also available on short-form video platforms such as TikTok, but users are free to choose. Furthermore, content on short-form video platforms is generally easily replicated, as the interfaces offer functions that allow, for example, users to use other users' videos as the basis for their own (Kaye, Zeng, and Wikström 2022, 23).

TikTok's predecessors include the short-form video platforms Vine, Snapchat, Flipagram, Musical.ly, and Douyin (for a chronological overview of the development of short-form video platforms, see Kaye, Zeng, and Wikström 2022, 27). The first internationally successful short-form video platform was Vine, which was acquired by Twitter (now X) in 2012 and officially launched a year later. The videos, which had a maximum duration of six seconds, were presented to users in the form of a feed, similar to the one that remains a fundamental aspect of X's current functionality. In the third year following its launch, Vine had over 200 million active users. However, in 2016, Twitter managers made the decision to discontinue the platform (Kaye, Zeng, and Wikström 2022, 34). The first half of the 2010s also saw the development of a phenomenon known as *stories*, which is not congruent with the short-form video format, but which helped shape the trend toward short media forms or *microformats* (Moormann et al. 2021). Stories are defined by their ephemeral nature, as they are removed from platforms after a designated period of time and can only be accessed once by individual users. The concept of the story was first introduced on the social media platform Snapchat in 2013. Snapchat is, in essence, a messaging platform. However, users were permitted to send one another what are known as *snaps*, which are photos or videos of up to ten seconds in length that are automatically deleted immediately after being viewed by the recipient. In 2016, Instagram introduced a feature called *stories*, which are automatically deleted after 24 hours. This feature has been a core aspect of the platform for several years (Kaye, Zeng, and Wikström, 2022, 35–37). A comparable feature has been offered by the platform Flipagram since 2013. It permits the creation of image slideshows of up to 30 seconds, with the option of adding music (Kaye, Zeng, and Wikström 2022, 38).

Musical.ly represents the principal predecessor of TikTok, and it possesses certain features that remain integral to the functionality of TikTok in the present day. Musical.ly was first launched in August 2014, following the completion of its development in China. From its inception, the stated objective was to establish Musical.ly as a successful platform in the U.S. market; the platform subsequently became popular in Europe as well. Musical.ly was originally conceived as a platform for sharing educational content. However, as this project was not successful, the focus soon shifted to music creation (Kaye, Zeng, and Wikström 2022, 40–42). It would be erroneous to assume that Musical.ly is the sole short-form video platform developed by Chinese companies.

Similar platforms have been highly popular in China since the early 2010s and serve as a prime example of Chinese web services that are specifically designed for international markets. For example, the company Kuaishou was established in 2011 and developed the short-form video platform of the same name, which remains highly popular to this day (Kaye, Zeng, and Wikström 2022, 25). As a consequence of the political climate in China, an autonomous platform ecosystem has emerged. International web services frequently face rigorous censorship by the Chinese government, which has prompted Chinese companies to develop their own platforms based on Western models and adapt them to the country's specific political circumstances (Kaye, Zeng, and Wikström 2022, 24).

The Chinese company ByteDance was responsible for the development of both TikTok and its sister platform, Douyin. Douyin was developed for the Chinese market, and TikTok is its international counterpart. With regard to functionality, the two platforms exhibit at most incremental differences (Kaye, Chen, and Zeng 2021). It is notable that Douyin was launched in 2016 – relatively late, given the developments in the Chinese short-form video platform market described above. In its inaugural year, the platform already had approximately 100 million users (Kaye, Zeng, and Wikström 2022, 48). In November 2017, Musical.ly was acquired by ByteDance and initially continued to operate under this name, followed by the launch of TikTok a few months later (Kaye, Zeng, and Wikström 2022, 43). Although no official figures are available, it is estimated that ByteDance invested between $800 million and $1 billion in Musical.ly (Kaye, Zeng, and Wikström 2022, 50). In retrospect, this acquisition can be seen as a strategic move that was well-considered and executed with a high degree of probability of success. For one thing, ByteDance had the necessary technologies to successfully develop and operate short-form video platforms, especially the algorithmically moderated video feed that was already a core feature of Douyin. In addition, the Chinese company now owned Musical.ly, a platform that had already achieved international success and was tailored to the interests of young users in the Western market (Kaye, Zeng, and Wikström 2022, 50). Musical.ly and TikTok were finally merged in 2018, continuing to operate under the name TikTok (Kaye, Zeng, and Wikström 2022, 44). The remarkable international success of Musical.ly and TikTok subsequently prompted other Chinese companies to develop short-form video platforms for markets outside of China. For instance, Transsnet created the platform Vskit for users in various African countries, while Kwai (based on the platform Kuaishou) was developed for the Brazilian market (Kaye, Zeng, and Wikström 2022, 26).

4.2 TikTok: Rise and Target Group

On a global scale, TikTok is one of the most successful and fastest-growing digital platforms of last few years. By 2020, two years after its official launch and merger with Musical.ly, the platform had already amassed 700 million active users outside of China. The number of monthly downloads reached its highest point during the initial global lockdowns that were implemented in response to the outbreak of the novel coronavirus. Over 100 million monthly downloads were recorded, and the TikTok app was downloaded over 300 million times in the first quarter of 2020. By February 2021, the total number of downloads since its launch had already exceeded 2.5 billion (Bhandari and Bimo 2022, 1; Kaye, Zeng, and Wikström 2022, 8).

As asserted by prominent developers and staff at ByteDance, TikTok and Musical.ly were designed with a predominantly youthful demographic in mind (Zeng, Abidin, and Schäfer 2021, 3162). In a public announcement, Alex Zhu, the founder of Musical.ly, stated that he had observed the online activities of younger consumers and discerned that their virtual pursuits were largely centered around music, dance, and online gaming. Social interaction with their peers was also of paramount importance. In order to address these interests and the need for social interaction, platform features were developed that allowed for low-threshold exchange while focusing on music-related content (Savic 2021, 3184). At the time of its inception, TikTok was predominantly utilized by a younger demographic. In 2020, for instance, 40% of TikTok users were between the ages of eighteen and twenty-four, while in the U.S. over 30% of the nearly fifty million daily users were fourteen or younger. One year later, the platform was being utilized on a daily basis by approximately 60% of the U.S. population between the ages of eighteen and twenty-four. However, the target group became significantly more diverse during the global lockdowns (Kaye, Zeng, and Wikström 2022, 171; Zeng and Kaye 2022, 80). The growth in the number of users demonstrates that TikTok experienced a notable surge in popularity, particularly during the first global lockdowns. During this period, the platform positioned itself as a source of entertainment and learning during challenging times of social isolation (Zeng, Abidin, and Schäfer 2021, 3163). Additionally, research findings indicate that the platform was utilized by numerous young individuals to process their experiences of the pandemic and share their perspectives with others (Klug, Kaufman, and Evans 2023, 153). TikTok served as a crucial coping tool for individuals grappling with mental health challenges stemming from social isolation. Additionally, the platform's array of entertaining content, despite its occasional frivolity, provided a much-needed source of distraction and enjoyment during a challenging period (Klug, Kaufman, and Evans 2023, 164).

4.3 Mimetic Practices

TikTok offers a plethora of features that are, at least in part, constitutive of various other short-video platforms. The "social media platform features" (Kaye, Zeng, and Wikström 2022, 72) typically offered by TikTok include functions such as liking, commenting, sharing content, and mentioning other users. Mentioning occurs via an @-symbol in the comment sections or video descriptions, which are linked to the user's profile. The use of hashtags is also prevalent; videos are typically tagged with multiple hashtags. Selecting a hashtag generates a list of all videos that utilize that hashtag, arranged in descending order of popularity (Kaye, Zeng, and Wikström 2022, 73–74). Additionally, TikTok provides a live streaming feature and an assortment of video effects and filters, which are accessible to the user via a categorized list in the control panel of the in-app video production tools (Kaye, Zeng, and Wikström 2022, 71).

Although all of these features are integral to common TikTok usage practices, it is the various interaction-enhancing features that distinguish TikTok from other short-form video platforms. Interaction with other users is a fundamental aspect of media platforms, including TikTok. On TikTok, however, the focus is not necessarily on interacting with one's own followers or friends. Rather, the focus is on the creative editing of videos by users who are not necessarily connected on the platform (in contrast to, for example, Facebook, where users are required to establish a connection). TikTok employs specific "platform features" (Kaye 2022, 60) to encourage editing of existing videos, thereby facilitating the creation of new content. In this sense, the platform offers a clear affordance aimed at the co-creation of audiovisual content.

The *duet* feature represents a fundamental aspect of this process. An existing video is integrated into a newly recorded video, resulting in a synchronous presentation of both videos in a split-screen format. A similar feature was already available on the Smule app, which was launched on iOS in 2012, and Musical.ly, TikTok's predecessor, also had a duet feature. However, it was not possible to record new audio tracks as part of a duet on Musical.ly. The feature was designed for lip-syncing and dance videos, and recording additional audio tracks was thus neither intended nor necessary. The duet feature was added to TikTok in 2019 and updated a year later to allow users to add their own audio tracks (Kaye, Zeng, and Wikström 2022, 65).

It can be argued that creative mimetic practices, in which users build on existing content from other users in individual ways, are central to the functioning of TikTok. The platform initially gained prominence for lip-syncing videos and dance challenges, which are based on the creative imitation or editing of existing TikTok videos (Kaye et al. 2021, 3196–97). The *stitch* and *use this sound* features are also integral to mimetic practices. The stitch feature enables users to extract a segment from an existing video and insert it at the beginning of their own video, maintaining both the original video and audio track, as well as the visual effects and on-screen text. TikTok users can ascertain which video a given audio track was initially used in and the number of subsequent videos created by other users based on that track (Kaye, Zeng, and Wikström 2022, 66). Another feature designed for interaction with other users is *video reply to comments*, which was officially launched in June 2020. This feature enables TikTok users to record new videos in response to comments made by other users in the comments section of one of their videos. The corresponding comment is then displayed in the video window of the newly produced clip (Kaye, Zeng, and Wikström 2022, 67).

4.4 Algorithmic Curation: The For You Page

Another key feature of TikTok that underscores the importance of algorithmic moderation to the platform is the *For You page*. This is the landing page that is automatically displayed when the app is opened on a smartphone. On this landing page, each user is presented with an algorithmically curated and individually personalized selection of videos. In this way, users are theoretically offered an "endless scroll of new videos" (Kaye, Zeng, and Wikström 2022, 58), because if they don't like a video, they can simply swipe it away and the next one is immediately displayed. Musical.ly already had a similar scroll feed, and it was one of the first features to be adopted on Douyin based on the Musical.ly model (Kaye, Zeng, and Wikström 2022, 51).

It is not clear to TikTok users why they are seeing certain content on their For You pages. TikTok's website does explain how the platform's algorithms work, but only on a very superficial level. According to TikTok, there are three factors that determine which videos appear in a user's feed: individual usage patterns, information about the videos a user has watched, and data about the user's devices. TikTok thus collects information about a user's interactions, such as which videos they have liked, shared, and commented on, and which accounts the user follows. In the video information category, TikTok collects information on which sounds and hashtags are used. And for devices, this information includes the device type, language settings or geolocation information (Kaye, Zeng, and Wikström 2022, 58; Bhandari and Bimo 2022, 5). In addition, according to official information from the company ByteDance, natural language processing techniques are used to classify text and audio elements, and computer vision technology is used to classify (audio)visual objects (Klug et al. 2021, 85). Taken together, it can be assumed that all of this information is used to present TikTok users with individually personalized video feeds that best match their personal interests. Because the specific workings of TikTok's algorithmic system (and digital platforms in general) are trade secrets, this information is admittedly not very specific.

While on platforms like Facebook, X, and Instagram, users' personal network of friends and followers has a significant impact on what content is shown to them, TikTok's focus is on algorithmic filtering that is largely independent of such networks. When the app is opened, the For You feed takes up almost the entire smartphone screen, while the buttons for news, comments and other functions are much smaller and relegated to the edge (Bhandari and Bimo 2022, 5). The user experience is, or at least appears to be, inherently dependent on the For You page and the corresponding algorithmic filtering logic. While algorithmic moderation is important for digital platforms in general, albeit as only one element among many, TikTok stands out from other platforms in that it is the core element in terms of user experience. Bandhari and Bimo state: "Of the major social media platforms on the market, TikTok is the only one to position its algorithm at the center of the social experience it engenders; the algorithm determines the type of video content the user is exposed to, and viewing this content makes up the majority of the experience on the platform" (Bhandari and Bimo 2022, 2).

Empirical studies of how the platform is used show that TikTok users consider the For You page and the algorithmic moderation that underpins this landing page to be a key feature of the platform. For example, several members of the JazzTok community emphasize that the For You algorithm is unique and that this is what differentiates TikTok from other (short-form video) platforms (Kaye, Zeng, and Wikström 2022, 112). Based on interviews with U.S. college students, Bandhari and Bimo show that TikTok users particularly appreciate the fact that the For You feed suggests content that actually matches their own interests and meets their individual quality standards. This provides a more satisfying user experience than, for example, the video streaming platform Netflix, whose recommendations one of the interviewees reported as generally not matching their actual interests (Bhandari and Bimo 2022, 7).

Observations of the TikTok user experience show that the adaptation of algorithmic recommendations to individual user interests is unusually fast. The TikTok algorithm learns quickly, which means that after a short period of time, only very specific content is displayed to individual users. This homogenization of content tends to be viewed positively by users, at least according to Bhandari and Bimo's findings. In particular, the accuracy of TikTok's algorithm has led some of its users to remain active on the platform over the long term. At the same time, several interviewees stated that some variety in the suggested content would certainly be desirable (Bhandari and Bimo 2022, 6). When it comes to algorithmic recommendations, there is also a tendency to anthropomorphize the TikTok algorithm, so to speak. This is particularly evident when respondents report experiences where they believed that the algorithm had understood them and the suggested content was highly relevant to their individual needs (Bhandari and Bimo 2022, 5). This is an example of a relatively close connection between human and non-human actors in digital spaces. The functional logics of the For You page thus produce a specific form of algorithmic culture in which users enter into a very close and intimate relationship with the platform.

Of course, it is typical of algorithmic cultures in the platform context that users do not know exactly which criteria are used to display or hide certain content – as is the case with TikTok. Because the use of TikTok is so strongly oriented toward engagement with the platform's algorithms due to the functional logic of the For You page, the platform literally induces the development of algorithmic imaginaries. Countless theories about how the algorithm works circulate among TikTok users, and one of the core concerns shared by many users is how to train the algorithm in favor of their own individual interests. For example, some users report experimenting with hashtags to understand how they can be used to increase their own visibility on the platform, without being able to develop reliable concepts. Other users monitored their For You pages for several months without posting anything themselves. This was done with the goal of identifying specific patterns in the use of hashtags and, more generally, recipes for success (Kaye, Zeng, and Wikström 2022, 62–63). There is also a widespread assumption among TikTok users that any form of interaction on the platform can help increase one's visibility. This means, for example, that users are constantly engaged in liking, commenting, and sharing content, and that content creators encourage other users to not only passively watch their videos but also actively respond to them using the available features (comments, etc.). Such hypotheses are also discussed in journalistic discourse. For example, there is an assumption that new videos are initially only recommended to a small group of users and, if they trigger lively interactions from users, are then further disseminated and displayed to more and more users on their For You pages (Kaye, Zeng, and Wikström 2022, 77). Other hypotheses posit that posting videos at certain times may lead to higher viewership. In an empirical study, Klug et al. compare such assumptions made by TikTok users with corresponding data on trending videos that TikTok makes available via an API (Klug et al. 2021). Among other things, they show that users believe that a large number of comments can contribute to the success of videos, which is reflected in the number of likes, shares, and views. The available data shows that such correlations do indeed exist, and that posting videos at certain times of day also seems to have a positive effect, while on the other hand no correlation can be established between the use of certain hashtags and the success of TikTok videos. However, it remains unclear whether, for example, a video has become successful because of the comments or whether many users comment on it because it was already popular beforehand (Klug et al. 2021, 89–91). Although users cannot be sure about the algorithmic logic, it is clear that hypotheses about what platform algorithms *might possibly* do can develop a power that should not be underestimated – after all, the behavior of TikTok users is to some extent guided by these algorithmic imaginaries.

4.5 Platform Affordances and Vernaculars

Due to the features of the platform that promote various actions (duet, etc.) and the core role played by the platform algorithm, TikTok strongly elicits specific user behaviors. As the cited empirical studies show, the platform's affordances have a strong impact on users, often leading to platform-specific behavior. This can in turn lead to the development of platform-specific norms of content production, i.e., make the creation of relatively homogeneous content more likely. Zeng and Abidin call this process *circumscribed creativity*: "Circumscribed creativities are creative potentials that are afforded, as well as restrained, by the features and logics of platforms. In the context of TikTok, the platform offers a technological infrastructure that celebrates and encourages memetic content" (Zeng and Abidin 2021, 2462). Examples of such mimetic content include the lip-synching and dance challenge videos described above, where it is not necessary for users to create original content – the challenge is rather to respond creatively to content that already exists on the platform. Thus, TikTok enables and provokes certain creative practices of imitation, while other practices that focus on the production of original content are initially less likely, at least in the context of lip-synching and dance challenge videos. It is also likely that the content of particularly high-reach content creators will appear on the For You pages of many TikTok users. When these videos are subsequently used by many users as the basis for new videos, the result is the production of countless videos with similar audiovisual content (Abidin and Kaye 2021, 59–60). One possible result of the platform logic described above is the production of large amounts of "aesthetically similar content" (Kaye, Zeng, and Wikström 2022, 79). It is therefore not surprising that some "platform-centric genres" (Southerton 2021, 3260) – such as the aforementioned lip-synching and duet videos – that have emerged on TikTok are also generally comparatively easy to identify as TikTok videos and would be unusual in their specific form on other platforms. Overall, it can be seen that TikTok promotes the development of specific platform vernaculars to a greater extent than other platforms due to the functional logics described above.

4.6 Niches on TikTok

While in many ways content production is driven by the logic of the platform – at least to some extent – and therefore may seem limited to certain types of content, TikTok also offers a wide range of topics. The platform may be best known for silly content by and for young people, but the content is diverse and there are various thematic niches on TikTok with specialized communities sometimes forming around niche topics. These are often related to education or science communication. For example, Southerton shows that videos created by medical professionals providing medical-related content are extremely popular on the platform. An investigation in March 2021 showed that videos tagged with the hashtag #doctorsoftiktok had a total of 428 million views at that time, while videos tagged with the hashtag #nursesoftiktok had as many as 1.8 billion views (Southerton 2021, 3250). Zeng, Schäfer, and Allgaier show that videos focused on science communication are extremely popular on TikTok and that presenting such topics is also compatible with the logic of the platform. However, this requires a specific framing of the scientific content, including in particular targeted juxtaposition with humorous elements (Zeng, Schäfer, and Allgaier 2021, 3230).

Young people's political awareness, for example on the topic of climate change, is also documented in countless TikTok videos (Basch, Yalamanchili, and Fera 2022; Hautea et al. 2021). According to Pomerantz and Field, TikTok has even become the platform par excellence for political activism among young people. Many TikTok users claim that videos with political content shared on the platform have helped them to learn about politics and social justice issues and to develop political awareness (Pomerantz and Field 2022, 63). By focusing on the sharing and editing of existing content, TikTok's functional logic can also foster issue-centered communities, which is why the platform is often used for the political activation and mobilization of young people (Sadler 2022, 4). Clearly, the functional logics or affordances of TikTok also influence how young people articulate and communicate their political attitudes and messages. For example, Zeng and Abidin show that the duet feature in particular allows TikTok users to directly reference other users' political statements, articulate approval or disapproval, and express alternative opinions (Zeng and Abidin 2021, 18). In this way, the platform ensures a low threshold for the articulation of political positions, which presumably makes it easier for young people in particular to access political debates.

4.7 Controversies

TikTok has been controversial from the start for two main reasons. The first is the potential influence of the Chinese government on the operations of the platform company ByteDance and thus how data trails generated by users are handled. The second is content moderation on TikTok, which is often subject to criticism. This is based on the assumption that certain content and certain groups of people can achieve greater visibility on the platform than others. As a result, the main critique goes, well-known cultural hegemonies based on categories of social difference such as gender and *race*, as well as normative notions of beauty, are perpetuated on TikTok.

The digital platforms that dominate the platform market in the Western world – Facebook, YouTube, Instagram, and X, for example – are run out of Silicon Valley. Although China has a large and differentiated platform market, especially in the short-form video platform segment, TikTok is a special case in that other platforms developed and operated in China are typically unable to compete with established platforms in the Western market. Much of the skepticism about TikTok stems from the fact that internet companies in China act as part of the Chinese government's surveillance apparatus. If the government suspects problematic content on the platforms, the operators must provide access to the relevant user data. ByteDance repeatedly claims to act independently of the government, but users and politicians remain skeptical. For example, TikTok was banned in some countries very soon after its launch (Kaye, Zeng, and Wikström 2022, 174–76). In this context, it is clear that internationally operating platform companies face the challenge of complying with the heterogeneous regulations regarding permitted and prohibited content in different countries without completely changing the way the platforms function on different regional markets.

Over time, critics have raised suspicions that TikTok's content moderation deliberately ensures that certain topics reach only a few users, while problematic content is sometimes allowed. Moderation on digital platforms typically involves defining a range of content that is not allowed and can therefore be deleted if discovered. TikTok does not differ significantly from other platforms in this regard, prohibiting, for example, the sharing of pornographic, violent, and discriminatory content (Kaye, Zeng, and Wikström 2022, 172–73). However, the platform's algorithmic moderation is regularly criticized, as content that does not meet moderation guidelines continues to gain popularity. In recent years, videos with problematic content, involving for example cyberbullying, misogyny, racism, xenophobia, and anti-Semitism, are said to have become popular on several occasions (Radovanović 2022, 53). The proportion of so-called "hate videos" with anti-Semitic content is said to have increased by more than 900% within a few years since TikTok's market launch (Divon and Ebbrecht-Hartmann 2022, 48). There are also repeated accusations of shadow banning. This refers to the allegations that the content of certain, often marginalized groups is deliberately not popularized on the platform. For example, there is a journalistic discourse that suggests that TikTok deliberately suppresses content posted by groups that may be at risk of becoming victims of bullying on the platform, such as people with Down syndrome or so-called "facial disfigurement" (Kaye, Zeng, and Wikström 2022, 183). However, this is not a phenomenon unique to TikTok; similar allegations are regularly made about other platforms as well. It is extremely difficult to obtain reliable evidence on this issue, as it is not possible for outsiders to gain concrete insights into the actual algorithmic functional logic or the sphere of TikTok's developers. However, two articles by journalists are often cited in this context, both of which are allegedly based on leaked internal documents from TikTok employees (Biddle, Ribeiro, and Dias 2020; Köver and Reuter 2020). These documents allegedly show that TikTok deliberately suppresses content from users or groups of people who do not meet certain beauty standards or who are disabled. However, the documents cited as sources in these articles are merely screenshots of Excel spreadsheets whose origin cannot be traced. Despite the fact that these theories are based on extremely vague sources, the platform responds to this type of criticism and tries to improve its image and generate positive headlines. For example, TikTok portrays itself as a platform where educational content plays an important role, and which aims to make BIPoC (Black, Indigenous, People of Color) visible and provide an inclusive space for LGBTQ+ communities (Kaye, Zeng, and Wikström 2022, 186–89).

4.8 TikTok and Cultural Hegemonies

One way to examine the prioritization – or suppression – of certain groups of people on TikTok is to look at popularity spikes, i.e., the most successful content creators on the platform in general or in specific segments. TikTok does not publish official figures on the most successful creators and their followers, but it is certainly possible to gather relevant information using the API, and various online services offer such overviews. Kaye, Zeng, and Wikström have compiled a list of the fifty most popular TikTok creators based on such sources, including information on the number of followers and likes, nationality, and year of birth. The data was collected in mid-2021 and is therefore no longer current, but the basic trends in popularization are clear (Kaye, Zeng, and Wikström 2022, 94–96). Due to the dynamic nature of TikTok's development, it is only possible to provide snapshots.

The list shows that the average year of birth of the most successful content creators was 1998, i.e., the average age at the time of data collection was twenty-three, with the youngest being born in 2004. The majority of creators therefore fall into Generation Z, those born between 1997 and 2012. Some of the top fifty include so-called "established offline celebrities" (Kaye, Zeng, and Wikström 2022, 97), such as actors Will Smith and Dwayne "The Rock" Johnson, as well as extremely well-known popular music singers such as Jason Derulo and Billie Eilish. These people, of course, were highly successful before the rise of TikTok, while most of the creators represented in this list are so-called "short-video celebrities" (Kaye, Zeng, and Wikström 2022, 97) who actually became popular through their presence on TikTok or Musical.ly. Two thirds of the people in the top fifty are American. One possible explanation for the dominance of U.S. content creators is that ByteDance has a strong economic interest in targeting the large U.S. market (Kaye, Zeng, and Wikström 2022, 97).

Most of the content creators in this list are *white*, which means that BIPoC are significantly outnumbered. In addition, the content creators who fall into this group are generally not short-video celebrities. While the second-ranked Black Italian creator, Khabane Lame, rose to fame through TikTok, other BIPoC creators, such as Will Smith, The Rock, and Selena Gomez, owe their popularity to their work in other segments of the entertainment industry. Typical examples of creators who have become well-known on TikTok include Charli D'Amelio, Addison Rae, and Noah Beck, who have gained attention primarily through the platform's signature lip-synching and dance videos (Kaye, Zeng, and Wikström 2022, 98). These individuals are *white* and, by normative standards, conventionally attractive "20-somethings" (Boffone 2022, 19), whose success, according to Trevor Boffone, is symptomatic of TikTok's popularization logic: "Notably, these TikTok influencers are all white or white-passing, reinforcing TikTok's racial power dynamics. While the app can be an inclusive, democratized space, TikTok fame mirrors fame more generally in the United States, which privileges whiteness and conventional Western standards of beauty" (Boffone 2022, 19). Boffone refers to the often-used argument, also promoted by TikTok itself, that the platform can, in principle, be an inclusive space. However, the field of popularity peaks is dominated by the so-called "D'Amelio aesthetic – slim, white and normatively attractive" (Ile 2021, 67) – this label being a reference to the most popular TikTok user at the time of data collection, Charli D'Amelio. This aesthetic may be particularly popular on TikTok because of the platform's mimetic logic. We can assume that the content of an enormously popular creator like Charli D'Amelio is displayed in the For You feeds of a large number of users, who in turn often respond to these videos by imitating the D'Amelio aesthetic, thus making their own success more likely and consequently perpetuating normative beauty ideals in terms of *whiteness*, thinness, symmetrical facial features, clothing, and makeup. This shows that there is an increased risk of perpetuating such cultural hegemonies through the logic of the TikTok platform.

There are now several empirical studies that demonstrate homogenization in terms of the perpetuation of certain beauty ideals, as well as heteronormative and racializing tendencies on TikTok, at least in certain segments of the platform. The fact that the most successful content creators are often young women who conform to the D'Amelio aesthetic was highlighted by Melanie Kennedy in a relatively early academic discussion of TikTok. Kennedy describes the "continuation and intensification of girl culture and the ideals of young female celebrities" (Kennedy 2020, 1071). In doing so, she emphasizes that TikTok to some extent perpetuates stereotypes of femininity that were already prevalent in society in general but may have been exacerbated on or by TikTok. As the author further argues: "It should not surprise us that the most-followed TikTok star is a slim, white, normatively attractive teenage girl (with straight white teeth, long straightened hair and her feminine body frequently displayed via tight fitting crop tops)" (Kennedy 2020, 1072). As Kennedy points out here, content creators display their slender, normatively beautiful bodies and serve as role models for countless, primarily young people who are active on TikTok.

Weight-related videos are also very popular on the platform. Minadeo and Pope show that hashtags such as #weightloss, #fatloss, #whatieatinaday, or #diet and related video content reach sometimes millions of users. Their sample of a total of 1,000 of the most popular videos in this segment is dominated by relatively young, *white*, college-aged people who primarily discuss weight loss and sometimes document their own weight loss process in their videos. The authors summarize: "This analysis found that nearly all of the 1,000 TikTok videos collected had content that was notably weight normative. Key themes included glorification of weight loss, the positioning of food to achieve health and thinness, and the lack of expert voices providing nutrition and health information" (Minadeo and Pope 2022, 9). This shows that even in this niche topic, which nevertheless reaches an enormous number of people, body stereotypes are extremely dominant. Once again, it can be assumed that more and more creators are turning to such videos due to the platform's virality-centric and imitation-oriented logic, and that the homogenization of content on TikTok is being driven forward by the platform logic.

Typically, such research focuses on young women and stereotypes associated with them. Foster and Baker, however, analyzed 200 videos from over forty of the most popular male creators on TikTok to examine masculine beauty ideals and stereotypical images of masculinity and male sexuality in digital spaces. They also found that people whose bodies do not conform to certain ideals of *whiteness* and thinness have a harder time competing for popularity on TikTok. The authors conclude:

> Importantly, TikTok's most widely followed men, as we have shown, embody a host of social privileges related to their race, youth, and physical appearance and possess masculine capital. Indeed, the overwhelming majority of TikTok's most widely followed men are white, cisgender, heterosexual, and physically fit and often muscular, providing visual evidence of their body work and effortful attention to appearance. (Foster and Baker 2022, 9)

According to Foster and Baker, male creators on TikTok certainly have the opportunity to consciously subvert these stereotypes, even if such content is more likely to reach a niche audience. This makes it clear that at least the most successful TikTok creators represent a binary gender logic. However, there are also copious amounts of videos circulating on the platform that challenge this binary, for example under the hashtag #trans. In an empirical study, Rochford and Palmer examined a total of 450 posts by eight successful trans TikTokers and concluded that such content can achieve significant reach. However, the most successful videos in this segment are tailored to the so-called cis audience, as this niche also caters to stereotypes that are likely to circulate primarily in circles that are not particularly familiar with trans aesthetics. This is expressed, for example, in the fact that the trans creators often use transphobic humor ironically or use TikTok's duet feature to respond to trans-hostile users. According to Rochford and Palmer, users who are familiar with the relevant codes and discourses may be particularly aware of the ironic connotations of such content (Rochford and Palmer 2022, 85–86). This shows that even in niche areas where stereotypes of binary gender images are actually being critiqued, the most popular videos are precisely those that are the most likely to meet the expectations of a cis audience.

The previous explanations also make it clear that success on TikTok is also influenced by intersectional logics. *White*, female creators who conform to certain normative ideals of beauty are generally more likely to succeed in the realm of popularity peaks (cf. Sweeney-Romero 2022). Black creators have complained for years that they are not able to gain as much traction on TikTok as their *white* counterparts. This has been the case, for example, with expressions of solidarity with the Black Lives Matters movement, and in this context several BIPoC creators have reported a significant drop in their views and likes (Peterson-Salahuddin 2024, 2391). In addition, *white* creators are often accused of appropriating the ideas of Black TikTokers and then enjoying greater success with them than the original creators (Ile 2021, 62; Martinez 2022, 40). In response, in 2021, BIPoC creators uploaded a series of videos under the hashtag #BlackTikTokStrike, declaring that they felt robbed of their ideas and no longer wanted to produce videos for TikTok (Ile 2021, 67–68). This leads to the suspicion that TikTok is also a "white space" (Ile 2021, 68), "fueled by an oppressive algorithm that privileges white, heteronormative content" (Ile 2021, 68).

4.9 Sound on TikTok

Due to the functional logic of TikTok, sound and specifically musical elements play a prominent role on the platform. In light of its predecessor platform Musical.ly, whose name emphasizes the importance of sound, this is obvious, and important platform features of TikTok are inconceivable without the use of music. After all, a key feature – *use this sound* – is based on the "spreadability of sounds" (Kaye et al. 2021, 3209), and on TikTok sounds often function as a mimetic medium, i.e. as an essential basis for the imitation practices that are central to lip-synching and dance challenge videos (Abidin and Kaye 2021, 59). In addition, musical elements are sometimes used to acoustically support the narrative of TikTok videos. For example, a pause in the action of a video may be accompanied by a musical drop. Sometimes musical elements are also used to support the "tonality" of a video, such as a shift from serious to sarcastic content (Abidin and Kaye 2021, 58). And thanks to the duet function, there is always the possibility of recontextualizing the auditory track. In this way, musical elements on TikTok contribute significantly to the dissemination of content and the creation of viral phenomena on the platform (Abidin and Kaye 2021, 61).

Since TikTok videos sometimes have an enormous reach and can therefore be used for marketing purposes, specific music-related processes of professionalization take place in the context of the platform. For example, short excerpts of songs – usually hooks – are uploaded to TikTok before the songs appear in full on other platforms such as Spotify. These excerpts are more likely to become viral phenomena, and it is assumed that professional music creators must learn to produce music specifically for use on TikTok (Radovanović 2022,

61). However, it is not only the music creators themselves who are undergoing a process of professionalization in this regard, but also people who specialize in supporting musicians in the context of TikTok. In recent years, a growing number of agencies have been founded that try to develop success strategies for musicians which they then communicate, for example, in the form of YouTube tutorials. Strategies for platform-compatible production and composition are also a recurring theme in this context; for example, the division of songs into approximately fifteen-second parts that can be used as the musical basis for dance challenges (Radovanović 2022, 62). In addition, TikTok is sometimes seen as an effective tool for jump-starting musicians' careers. For example, the phrase "TikTok-to-Spotify pipeline" refers to the possibility of being placed on major Spotify playlists as a result of becoming successful on TikTok and thus reaching a large number of listeners (Radovanović 2022, 63).

4.10 Revenues on TikTok

The TikTok-to-Spotify pipeline is important for musicians because TikTok has been used primarily as a marketing tool, rather than as a direct source of income for musicians. A viral video on TikTok will not automatically result in a large amount of revenue, but there are several ways for musicians to earn money directly on the platform. One way is through the live stream feature, which is only available to users who already have a large number of followers. Users can buy TikTok coins, which they can use to buy gifts and send them to creators as part of a live stream. The creators in turn have these gifts converted into diamonds, which are then converted into real money and paid out, with the platform retaining 50% (Kaye, Zeng, and Wikström 2022, 151). Sponsorship deals with companies who want their products promoted by TikTokers are another way for creators to generate revenue. This is a classic source of income in the influencer industry, and the TikTok Creator Marketplace was launched in the U.S. market in early 2020 to bring advertisers and content creators together. There is also the TikTok Creator Fund, which was launched in July 2020. TikTok is tight-lipped about the specific distribution mechanisms, but in principle, content creators are compensated individually and based on their level of success. Initially, only U.S. TikTokers were eligible to receive compensation from the Creator Fund. They also had to be at least eighteen years old, have 10,000 followers, and have 100,000 video views in the previous month (Kaye,

Zeng, and Wikström 2022, 154). Creators in several European countries are now also eligible to participate in the fund (TikTok n.d.; TikTok 2021a; TikTok 2021b).

Overall, it can be seen that the platform specifics of TikTok can have a significant impact on the practices of users who are active on the platform. TikTok differs from other platforms in several ways: First and foremost, the numerous features that elicit mimetic practices and the absolutely crucial role of the recommendation algorithm, which is at the heart of the user experience. The studies mentioned in this section show that TikTok users sometimes interact very intensively with the platform's algorithm. It is also clear that the homogenization of content is most likely due to the logic of the platform and users' interpretations of it. The principal question for this study is how these socio-technical relations affect the representation of jazz on TikTok.

5. Methodology

This study aims to identify the facets of jazz that are the most popular on TikTok. What affordances of the platform prompt users to engage with music-related actions? To what extent do musicians develop algorithmic imaginaries that guide their behavior on TikTok? It would be interesting to see whether jazz-specific platform vernaculars emerge on TikTok. To what extent do the functional logics of the platform, in interaction with the musicians who hope to become successful on TikTok, influence the representation of this music culture? It would also be interesting to see whether certain (groups of) musicians become more visible than others, and whether this perpetuates cultural hegemonies. Finally, it would be interesting to see what image of jazz is created on TikTok and for what reasons.

Two principal analytical procedures are required in order to address these questions. Initially, a compilation and examination of the most prevalent jazz-specific videos on TikTok must be conducted. This necessitates the delineation and organization of the overarching characteristics of the musicians and their performances, as well as the musical compositions that are played. This step is essential in order to illustrate which musicians and repertoires have the most extensive reach on the platform and consequently exert the most significant influence on the representation of jazz on TikTok. Furthermore, we can commence the process of interpreting which facets of jazz are more popular on TikTok than others, and for what reasons. Secondly, an in-depth examination of jazz musicians who are popular on TikTok is required to ascertain how these artists navigate the platform and, when necessary, adapt their musical output to align with its logic. It is thus essential to engage in dialog with musicians who are at the vanguard of their respective fields, as they can offer insights into the mechanisms of popularity on TikTok on the basis of their first-hand experience. Consequently, an inductive analytical approach is necessary, progressing from a comprehensive initial analysis to a more detailed and nuanced examination.

5.1 Corpus Analysis

To systematically examine the characteristics of popular jazz videos on TikTok, one effective method is the corpus analytical approach. This approach involves the systematic organization of content from multiple videos that are selected for being as representative as possible within a specific field of investigation. This content is organized on the basis of a set of pre-defined descriptive categories, which are modified or refined during the research process. These may include, for example, the musical repertoire played in the most popular jazz videos on TikTok, the settings in which musicians perform, or the assignment of musicians to categories of social difference, such as gender and *race*. In this way, certain patterns and regularities can be identified within a limited cultural field. At the same time, initial interpretations can be formulated regarding content and musical forms of expression that may be more relevant than others within the given field, and for what reasons. Such an exploration of the field of study is a particularly useful tool for research on digital platforms. The sheer volume of content on digital platforms necessitates a targeted and judicious selection of research cases. While this selection should not be arbitrary, it is also important to recognize the limitations of comprehensive analysis within the context of such a vast and diverse corpus of material. The objective is to conduct a rigorous and systematic exploration of the field, guided by a well-considered and well-founded selection of illustrative examples that are as representative as possible.

In order to make such a selection, it is first necessary to conduct an open and focused observation of jazz-related content on TikTok, preceding the actual corpus analysis. This is essential for the researcher to gain insight into the underlying logic of video popularity within the chosen field of investigation. It enables the identification of key content creators, repertoires, and representational conventions, as well as the formulation of criteria for a meaningful selection of examples for analysis. The *walkthrough method* (Light, Burgess and Duguay 2018) was developed for the systematic exploration of digital platforms. In this approach, the researcher becomes acquainted with the pertinent rules and functional logics from the perspective of the user through targeted observation and active use of the respective platform or its corresponding app:

"The walkthrough method is a way of engaging directly with an app's interface to examine its technological mechanisms and embedded cultural references to understand how it guides users and shapes their experiences" (Light, Burgess and Duguay 2018, 882). This systematic process commences with signing up on the digital platform as a user. This is followed by a series of phases of actual use, which conclude with the observation phase, which may end with deregistration or discontinuation of use. One central concern in this approach is to identify the affordances of the respective applications or platforms. Additionally, the aim is to understand why certain usage patterns are more likely to emerge on certain platforms than on others (Light, Burgess and Duguay 2018, 886). Despite the authors' proposal of a comprehensive analytical framework, it has become evident that a singular, inflexible methodology cannot be universally applicable to all research endeavors in the platform context. This is largely due to the inherent diversity and dynamism of platforms, which are constantly evolving (Duguay and Gold-Apel 2023, 1). Nevertheless, the systematic investigation of the selected field is a fundamental aspect in all cases, although the focus may vary depending on the subject matter and research question. For instance, several researchers employ the walkthrough method to analyze content on TikTok (see, for instance, Bhandari and Bimo 2022; Zulli and Zulli 2022). In some instances, comparable approaches are also referred to as *digital ethnography* (e.g., Schellewald 2021; Southerton 2021; see also Hjorth et al. 2017; Pink et al. 2016). Some of the authors utilized the TikTok platform for a period of several months, during which they documented their observations and organized the content in a systematic manner. For example, they sought out prevalent topics, hashtags, and creators, observed interactions between creators and other users, took screenshots, and downloaded videos with the aim of gradually becoming acquainted with the established norms within their respective field of study (cf. Schellewald 2021, 1441–42; Southerton 2021, 3253).

Following the observation phase, the next step is to select the cases for analysis in the corpus. In order to ensure that the selected analysis cases are as representative as possible within the chosen field of investigation, it is essential to implement a considered and strategic selection process. Selecting videos based on recommendations on the researcher's For You page is not an appropriate approach. While observing TikTok activity, it is essential not to overlook the landing page as a crucial element of the platform. However, as previously discussed, the suggestions on the For You page are generated through automated analysis of individual user behavior, making it unlikely to yield insights into the content that is particularly popular among other users and therefore representative of a specific field of research.

If the objective is to identify peaks of popularity within a defined research area, using hashtags is a logical approach for locating suitable cases for analysis. The non-mandatory but common practice of using hashtags serves to categorize videos on TikTok to a certain extent by linking them to specific keywords and assigning them to particular topics. During the observation phase, it is essential to identify which hashtags are popular within the respective field of investigation and could, therefore, be relevant for the selection of cases for analysis. A compilation of videos pertaining to specific topics on the platform can then be created based on the use of designated hashtags. This is a logical conclusion when analyzing popularity peaks within a limited field of study. The objective is to understand which type of content or creator is most likely to succeed on the platform (Hautea et al. 2021, 5). In the extant literature on TikTok, it is customary to commence analysis with the most popular videos, such as those with the highest viewership or like count tagged with a specific hashtag (e.g., Minadeo and Pope 2022, 3; Ryan and Tileva 2022, 168–169; Southerton 2021, 3253).

Once the pertinent hashtags have been identified, web scrapers can facilitate the aggregation of data. This software is capable of extracting and analyzing data provided by platforms through application programming interfaces (APIs). Web scrapers can be utilized to collect and organize information on a multitude of videos. To illustrate, if one is searching for a particular hashtag on TikTok, web scrapers enable the compilation of all videos that have been tagged with that hashtag on the platform. Additionally, the number of likes, views and comments, the time of posting, and the length of the video are often provided as supplementary information. The metadata associated with TikTok videos can be organized and prepared for further analysis using standard software, such as spreadsheets. The use of web scrapers is common practice in TikTok research, particularly in the context of studies that require the analysis of large data sets. For example, Klug et al. collected and analyzed data from more than 300,000 videos (Klug et al. 2021, 86; see also Hautea et al. 2021, 4; Zeng and Abidin 2021, 5; Zeng, Schäfer and Allgaier 2021, 3220–22). It should be noted that only data made available by TikTok via the application programming interface (API) can be collected in this manner; in-depth information on the specific usage patterns of individual users or even the underlying functional logic of the platform is not accessible. As Jean Burgess notes, the "research affordances" (Burgess 2021, 31) of digital platforms present a significant challenge for researchers. The functional logics of these platforms often exert a decisive influence on the types of research questions that can be posed within the platform context. Consequently, platform-related research is, to a certain degree, inherently constrained by the platforms themselves.

Once the cases for analysis have been selected, the content of the TikTok videos must be systematized. This entails assigning the videos to a variety of descriptive categories to identify specific patterns and regularities. Similar approaches have long been employed in the field of image analysis (cf. Bell and Milic 2002) and are suitable for studying large corpora of Instagram images (e.g., Lucibello et al. 2021; Zappavigna 2016). In the context of research on TikTok, corpus analytical approaches are frequently employed to categorize the content of multiple videos within a specific research area (see e.g., Schellewald 2021; Zeng and Abidin 2021; Zeng, Schäfer and Allgaier 2021). For the analysis, appropriate categories must first be defined to which the individual elements of the selected videos can be assigned. The system of descriptive categories developed in this step must be further refined throughout the research process in order to adequately categorize the heterogeneous audiovisual content. While there are no binding guidelines on which analytical categories to apply, the analytical procedure must always be concretized in line with the specifics of the respective field of analysis and the individual research question. This is necessary, in part, due to the heterogeneity of the potential analysis examples. For example, in the context of more general observations on recurring patterns of TikTok videos, the focus can be on categories such as the type of protagonist (e.g., ensemble or solo artist), the setting (e.g., bandstand, rehearsal room, or bedroom), and the visual effects (e.g., filters) (Schellewald 2021, 1443). A focus on the systematization of TikTok-specific practices is also conceivable, for example with regard to the sounds used in a video (original or template) or the performative practices that are typical of the platform (such as lip-synching or dance challenges) (Zeng and Abidin 2021, 6). If the focus is on a specific topic, such as science communication on TikTok, it is recommended that the videos be organized according to categories such as the respective scientific discipline (e.g., chemistry, physics) or the method of knowledge transfer in the videos (e.g., demonstration of experiments or verbal explanation) (Zeng, Schäfer and Allgaier 2021, 3223–25).

5.2 Undertaking the Corpus Analysis

In line with the key work steps described above, the first step was to explore and observe the field of jazz on TikTok. Once the core idea of investigating the representation of jazz on TikTok had been established, the exploration commenced with preliminary Google searches, exemplified by the search term combination "TikTok jazz." The results at the top of the list pertained to a number of prominent creators whose usernames included the term "jazz," yet who were largely not associated with jazz in a musical capacity. However, the most frequently occurring search results invariably included a link to the page in the TikTok desktop view, situated under the heading "#jazz | TikTok." This page lists videos that have been tagged with the hashtag #jazz, sorted by popularity in terms of views and likes. By visiting this page and viewing a substantial number of the listed videos, we were able to gain insight into the specific aspects of jazz that are particularly popular on the platform. The TikTok application was then downloaded onto a smartphone to facilitate a user-centric examination of the conventional platform. Concurrently, popular jazz creators and popular jazz-specific hashtags were searched for on both Google and the TikTok desktop and mobile applications. The results indicated that the #jazz hashtag was the most prevalent hashtag with a distinct jazz reference on TikTok. At the time of the initial research on this topic in early 2022, videos tagged with this hashtag had been viewed over two billion times in total. Approximately two years later, this number had nearly quadrupled. Other popular hashtags included #jazztok, #jazzsinger, #jazzstandard, and #greatamericansongbook.

In terms of content, it became evident that younger individuals on TikTok tend to achieve notable success by performing relatively old U.S. jazz standards, in alignment with hashtags such as #jazzstandard and #greatamericansongbook. While improvisation – a hallmark characteristic of jazz – is not a prominent feature in the majority of the videos observed during this phase, a considerable number of videos with a comedic focus were identified. The videos are typically tagged with hashtags such as #jazzmemes and employ a jocular tone with regard to conventional jazz stereotypes. These include the use of specific vocabulary and musical clichés that are pervasive in the jazz genre. Videos featuring the rhythmic-melodic pattern known as *the lick* are particularly prevalent. This pattern occurs frequently in jazz improvisations and has gained recognition within the jazz community due to a compilation of such passages that was posted on YouTube (Judd 2022; see chapter 2). Similarly, interpretations of John Coltrane's "Giant Steps" (Spencer 2023) are also popular, often with an ironic twist. In terms of reach, however, this content is significantly less popular than the most popular videos distributed under the #jazz hashtag, which is why I decided to focus on videos tagged with that hashtag and the musicians associated with it. Focusing on this hashtag also allows for the examination of content that is presented as jazz-specific on TikTok, providing a more objective analysis of the facets of jazz that are the most popular on the platform.

Using a web scraper (Apify 2024), I compiled a list of the one hundred most popular, i.e., most viewed, videos tagged with the hashtag #jazz, including metadata. A total of one hundred videos were subjected to a detailed content analysis, with the aim of providing an interpretative basis for further, more detailed analyses (cf. Eriksson Krutrök 2021). A small number of videos lacked any sonic or visual references to jazz as a musical genre and were therefore excluded from the analysis. In such cases, the next most popular videos were added to the list until the target number of one hundred videos was reached. This approach allowed me to identify some of the most popular jazz-related TikTok videos overall. The primary limitation of this methodology lies in the fact that some musicians do not consistently utilize this specific hashtag, even though their videos potentially have a clear connection to jazz. To ensure that the selected videos and musicians were representative of the broader jazz landscape on TikTok, the top one hundred videos were randomly compared to those shared under other popular hashtags, including #jazztok and #jazzsinger. Given the absence of any significant discrepancies in terms of repertoire, production style, or the (groups of) musicians featured in the videos, an exhaustive examination of other hashtags was deemed unnecessary. Throughout the process, I conducted regular assessments of the top one hundred videos tagged with #jazz to identify any notable shifts over time. My findings revealed that new videos were only sporadically identified within this list and did not diverge substantially from the videos selected for the corpus analysis in terms of content or the musicians' characteristics. The popularity peaks in the field of jazz on TikTok appear to exhibit a certain degree of constancy. Consequently, the corpus analysis was ultimately limited to the one hundred initially selected videos. It is, however, possible that the selection process may have resulted in the exclusion of some videos and musicians that are in fact part of the popularity peaks in the jazz field on TikTok and should therefore be included in this study. However, due to the vast quantity of jazz-related content on TikTok, it is unavoidable that a targeted reduction of potential examples for analysis will be necessary. It is important to critically reflect on the possible disadvantages of the chosen approach, but this must ultimately be accepted for methodological and practical reasons.

At the outset of the corpus analysis, each video was downloaded and viewed multiple times. Any regularities regarding the sound or visual aesthetics, the musicians, and TikTok-typical features (duet, etc.) were noted in the form of memos as soon as they were identified. These steps were necessary in order to inductively develop a set of categories from the material, on the basis of which the content of the videos could later be systematized. The content analysis was conducted using MAXQDA software, which is designed for the analysis of qualitative data. While this software is often utilized for the analysis of textual data, such as transcripts of interviews and journalistic articles, it can also be used to analyze (audio-)visual data. The first step is to define the analytical categories, after which the data – in this case the videos – is coded, i.e., specific content elements are assigned to particular thematic categories. MAXQDA offers the option of displaying all the videos that have been assigned to a certain category in a list, thus facilitating a comparison of videos assigned to a given category. This allows for the refinement of the category system as needed during the analytical process. The content analysis resulted in a category system comprising approximately 1,700 codes. On the basis of my own observations and the corpus analytical studies on TikTok cultures discussed above, I identified the following main categories, which are themselves comprised of numerous subcategories and sub-subcategories: musical repertoire, gender relations, *race*/ethnicity, setting, video form, musical performance, and verbal elements. It is important to note that categorizations based on social categories such as gender and *race*/ethnicity are not entirely reliable and can only be made in a relatively superficial manner within the context of such an analytical method. For example, the classification of musicians as female or male was based on my own subjective perception, which is influenced by the pervasive logic of binary gender categorization (Minadeo and Pope 2022, 4). It is possible that the actual gender identity of the individuals depicted in the videos may differ from these categorizations.

5.3 Results of the Corpus Analysis

5.3.1 Musical Repertoire

There is a clear tendency toward versions of jazz standards played by the content creators: this phenomenon was observed in forty-six out of one hundred videos. Most of these songs are part of the repertoire contained in the Great American Songbook, a loosely defined canon of popular songs composed and first recorded from the 1920s to the 1950s. These songs mainly stem from Broadway musicals, Hollywood films, and Tin Pan Alley (an early *hit factory*, see Seabrook 2015). A significant number of these compositions have been performed by jazz musicians over the years, and they represent a specific aspect of the jazz canon: these songs are considered *jazz standards* (cf. Michaelsen 2013; Williams 2023). In twenty-one videos, we can hear original compositions by the content creators, albeit not necessarily fully fleshed-out songs, but rather short loops or harmonic progressions. The musicians Stacey Ryan and Laufey are the only ones to perform snippets of original songs. In fifteen videos, the original sound of existing jazz recordings was adopted. While the adoption of pre-existing sounds may be typical for many TikTok videos, especially in the case of dance challenges and lip-synching, it is obviously of less importance for jazz musicians on the platform. Improvisation plays a minor role, as we hear musicians improvising in only eleven videos. Most of the time, these improvisations are only very short segments without instrumental accompaniment, while only one musician, the trumpeter Kellin Hanas, improvises to jazzy instrumentals in a classical sense.

Most of the time, we hear versions played by the content creators or original recordings of songs first recorded from the 1930s to the 1950s: there are ten songs from the 1950s and nine each from the 1940s and 1930s. The oldest song in the corpus is from 1928 ("Makin' Whoopee," performed by Laufey). Table 1 provides an overview of the songs that appear in the corpus at least twice.

Table 1: Songs that appear at least twice in the corpus.

Title	Songwriters	Year	Occurrences
"Fly Me to The Moon"	Bart Howard	1954	5
"It's Been a Long, Long Time"	Jule Styne/Sammy Cahn	1945	4
"In the Mood"	Wingy Manone/ Andy Razaf/Joe Garland	1939	3
"L-O-V-E"	Bert Kaempfert/ Milt Gabler	1964	2
"La Vie En Rose"	Édith Piaf/Louis Guglielmi	1947	2
"Sing, Sing, Sing (With a Swing)"	Louis Prima	1936	2
"Sway"	Luis Demetrio/ Norman Gimbel	1954	2
"It Don't Mean a Thing (If It Ain't Got that Swing)"	Duke Ellington/ Irving Mills	1932	2

However, it is not clear whether the musicians are really referring to the first recorded versions of the songs, or rather to the most popular recordings. For example, the song "Fly Me to the Moon" was first recorded in 1954 but was popularized by the version recorded by Frank Sinatra in 1964 – nowadays, it is widely considered a Sinatra song.

5.3.2 Gender Relations

The gender ratio is relatively balanced, with a slight tendency towards male musicians. Videos were coded as "male" or "female" if the musician who uploaded it can be assigned to this gender category. This is also true for duet videos, where two or more videos from different accounts are displayed simultaneously. In forty-seven videos, there are male musicians performing, and women perform in thirty-seven videos. In ten videos, we see mixed groups, and in six videos, the musicians' gender is not discernible as the musicians themselves are not visible.

The gender ratio might lead one to assume that the most popular jazz musicians on TikTok are male. If we take a closer look at the musicians who appear in the corpus multiple times and whose musical repertoire is typical with respect to the aforementioned focus on Great American Songbook songs from the first half of the twentieth century, however, the picture is slightly different. The musician featured most frequently in the corpus is Ricky Rosen (@rickyrosen), who appears in seven videos. He performs versions of songs such as "Sway" (originally from 1954) and "It's Been a Long, Long Time" (originally from 1945). Another male musician who performs a comparable repertoire is Erny Nunez (@young_crooner), with two videos in the corpus. Ben Freeman (@ben_makes_names_to_music) has five videos in the top one hundred: he provides humorous content by, for example, playing songs backward and asking the audience to identify the song they hear – this kind of jazz performance is rather unusual with regard to the analysis of the musical repertoire. The most successful female musicians on the list, on the other hand, clearly represent a strong focus on Great American Songbook songs. Stacey Ryan, Laufey, Ebony Loren, and Rachel Chiu each have at least two entries with versions of songs such as "Fly Me to the Moon" (1954) and "It Don't Mean a Thing (If It Ain't Got that Swing)" (1932). This means that the jazz repertoire that is the most popular on TikTok tends to be represented by female musicians who are highly popular on the platform, each with at least 500,000 followers. Although the musicians' age is not displayed on TikTok, all the musicians mentioned here are clearly in their early twenties or even younger.

5.3.3 *Race*/Ethnicity

The relationships regarding the *race*/ethnicity category are very clear. The coding followed the suggestion of Lucibello et al. and assigned the videos to the categories "Asian, Black, Indigenous, Latinx, Middle Eastern, White, Other" (Lucibello et al. 2021, 150), with two additional categories: "mixed groups" and "not discernible." The category "mixed groups" was chosen if the videos featured multiple musicians who could be assigned to different categories. If, for example, only musicians belonging to the category "Black" were shown, no distinction was made between individuals and groups. "Not discernible" was selected if the musicians were not recognizable, e.g., due to costumes. These categorizations, like those of gender, are inevitably superficial. However, in the context of a systematizing rough analysis, such an approach is unavoidable if we are to first reveal any inequality relations with regard to social categories of difference on a general level and thus lay the foundation for more in-depth analytical steps.

The defined top one hundred list is clearly dominated by *white* musicians; this category was assigned to a total of sixty-eight videos. The categories "Black" (5), "Latinx" (4), "Asian" (3), and "not discernible" (3) play a clearly subordinate role, while the categories "Indigenous," "Middle Eastern," and "other" were not assigned at all. Six videos do not show any people. This overview shows that the musicians who can be clearly categorized as *white* are clearly in the majority. Evidently, the domain of jazz popularity peaks on TikTok is predominantly a "white space" (Ile 2021, 68).

5.3.4 Setting

No great variability can be ascertained with regard to the video settings. In the majority of the videos (73), we see the musicians perform in domestic spaces. Only occasionally are they filmed playing on a stage or in a rehearsal room. This is typical for TikTok, as many creators film themselves at home – an aesthetic that was certainly influenced by the coronavirus pandemic, when a large part of the global population was in domestic isolation, and which continues a trend that has already been described as "'bedroom culture'" (Kennedy 2020, 1070) in social media research, even before the advent of TikTok. On TikTok, however, this type of staging is very common. This may be due to the fact that TikTok experienced its greatest surge in popularity during the first global lockdowns during the coronavirus pandemic, when countless people around the world were isolated in their homes and forced to record their videos in such settings.

5.3.5 Video Form

Most of the videos (74) are obviously filmed by the musicians themselves by placing their smartphone in front of them. Most musicians (85) are filmed in selfie mode, i.e., from a relatively close distance, while fifteen musicians are seen from a greater distance, for example performing on a stage. Even though the duet is one of TikTok's key features, only seven videos in the corpus make use of this feature. The majority of videos (73) are recorded in a single pass, which means that storytelling modes with more than one storyline or perspective do not play a key role for most musicians.

5.3.6 Musical Performance

In eighty-seven videos, we see musicians performing the music themselves, whereas in thirteen videos the music is taken from other sources. There is a clear tendency towards solo performances (62); ensembles are featured in twenty-five videos. This would seem to make sense, as the majority of videos are produced by musicians at home, not in a concert or rehearsal setting. In thirty-seven videos, we see musicians singing, optionally accompanied by an instrumental track (16) or their own instrumental accompaniment (12), a cappella (8), or accompanied by multiple instrumentalists (1). While there is no clear tendency regarding the musical accompaniment, singing along to pre-recorded instrumentals might be one feature that is typical for jazz performances on TikTok, but rather unusual in jazz outside the platform. Instrumental music without vocals occurs in thirty-two videos.

5.3.7 Verbal Elements

The majority of videos use the English language, be it in the song lyrics, the text layers, or spoken language. Spanish (4) and French (2) language elements occur occasionally, while ten videos contain no verbal elements at all. On TikTok, creators can add written text to their videos by using the app's features for text layers, for example to explain the topic of the video or display song lyrics. One reason for doing so might be that there is not enough time for spoken explanations or introductions due to the platform's short-video aesthetic. Furthermore, adding song lyrics can make it easier for other users to apply the duet feature, as they can sing along to the original sound. The text layers might also attract more user attention. In several videos (22), the text layer is used to explain what the video will be about. Song lyrics are displayed in fifteen videos, and in fourteen videos the creators use the text layer for telling stories that are not related to the topic of the video or the song lyrics. Song titles are displayed in twelve videos, and some (11) creators reply to users' comments, which are displayed in a text box on the left of the screen. This feature cannot be regarded as a text layer in the same way as the aforementioned examples, but it does add verbal elements to the videos. These textboxes typically display comments by users and requests for the creators to sing particular songs.

In twenty videos, we can hear spoken announcements by the musicians: the song they are about to sing, for example. In the videos with more humorous connotations, the announcements sometimes serve to set up the punchline of a joke (12).

5.4 Prototypical Contents and Musicians

Of course, there is no definitive formula for jazz performances on TikTok, although some typical patterns can be identified based on corpus analysis. Particularly popular are versions of jazz standards from the Great American Songbook repertoire, first recorded between the 1930s and 1950s and largely written by *white*, male American songwriters. On TikTok, these songs are usually sung and played by relatively young *white* women in their early twenties, although there are more male musicians in the defined top one hundred. Musicians tend to perform in a domestic setting, filming themselves in selfie mode, and vocal performances are more popular than instrumental music. Almost all musicians sing songs in English, which is obviously their native language. Text layers are regularly used, while verbal announcements are rather atypical, as long as the musical performance and not the humorous connotation is in the foreground.

According to the corpus analysis, the following musicians can be described as prototypical for jazz on TikTok, as they are represented several times in the top one hundred, singing comparatively old jazz standards, and having at least a six-digit number of followers, i.e. a comparatively large reach: Ricky Rosen (@rickyrosen), Stacey Ryan (@staceyryanmusic), Laufey (@laufey), Rachel Chiu (@rachelchiu1), Ebony Loren (@ebonylorenmusic), Sam Ambers (@sam_ambers), Stella Cole (@stellakcole) and Erny Nunez (@young_crooner). If we take a closer look at their profile pages, it becomes evident that the focus on singing jazz standards from the Great American Songbook repertoire dating from the early twentieth century to the 1950s is a defining factor for the style of their performances on TikTok. For example, almost all of them have uploaded their own version (sometimes even several versions) of the two songs that, according to the corpus analysis, are the most popular: "Fly Me to the Moon" (1954) and "It's Been a Long, Long Time" (1945). That is to say, the jazz musicians who are the most popular on TikTok also perform the songs that enjoy the greatest popularity on the platform. This is not only true for the two songs mentioned above, but also for "La Vie en Rose" (1947) and "L-O-V-E" (1964), for example. This gives the impression that popularity on TikTok also depends, at least to some extent, on the choice of a specific musical repertoire.

The specific facet of jazz represented by the aforementioned songs is not directly related to the canonized jazz repertoire or to musicians that have typically been presented as the standard for jazz in journalism, academia, and music documentaries for decades. The music of so-called "jazz icons" (Whyton 2010) such as Charlie Parker, Miles Davis and John Coltrane is completely absent. Songs such as "La Vie en Rose," a signature tune of the chanson singer Édith Piaf, and "L-O-V-E," composed by Bert Kaempfert and Milt Gabler, would probably not be described or recognized as jazz by many jazz fans and musicians. However, in the context of a short-form video platform that is highly interactive and encourages imitation, it is evident that such songs will gain popularity. It can be reasonably assumed that the majority of people would consider these songs to be more catchy, more singable, and more accessible than the music of jazz icons. It is likely that such songs are more conducive to attracting attention on TikTok than the virtuoso improvisations and harmonically complex compositions that are typically highly valued in jazz circles. These compositions often span several minutes and lack particularly catchy melodies. It can thus be posited that the functional logic or affordances of the platform exert a considerable influence on musicians' selection of musical repertoire.

Some of the most prominent jazz musicians on TikTok are relatively young *white* women who conform to conventional standards of beauty. While they occasionally perform solo on guitar or piano, it can be assumed that the majority of them are perceived as singers. The results of the corpus analysis indicate that jazz instrumentalists are typically less prominent on TikTok. It is unsurprising that stereotypes related to gender and *race* also play a significant role on TikTok, as digital platforms are suspected of perpetuating certain cultural hegemonies (see chapter 3.3). Therefore, it seems that not only the selection of a highly specific repertoire, but also a particular appearance and physiognomy is of paramount importance for success on TikTok.

The analysis of the corpus reveals a striking homogeneity in the representation of jazz on TikTok. It is notable that certain canonized aspects of jazz are also prominently represented on the platform. Most notably, this concerns a certain North American centricity typical of jazz discourses, largely due to the dominance of the Great American Songbook repertoire. Nevertheless, this may not be regarded as the definitive measure of jazz in academic circles. Concurrently, young female vocalists appear to be acquiring a notable degree of visibility, despite the fact that this group of musicians is frequently excluded from conventional jazz historical narratives. It would seem that a tension between the perpetuation and renegotiation of jazz-specific conventions and stereotypes is arising in the jazz popularity peaks on TikTok. In order to examine and contextualize these findings in greater depth, further qualitative empirical steps are necessary.

5.5 Interviews and Conversational Analysis

The corpus analysis facilitated the identification of the specific types of jazz-related content that are particularly popular on TikTok. However, the question of why this content is so successful on the platform remains unanswered. As outlined in chapter 3.7, qualitative research on the production logics of successful TikTok creators represents a suitable approach to gaining a deeper understanding of the platform-specific processes of popularization. In light of the secrecy that pervades platform companies and their sporadic bans on research projects that delve into specific platform logics (see Eriksson et al., 2019), it is logical to conclude that researchers cannot expect to acquire direct insights into the operational mechanics of algorithmic systems or the mechanisms through which specific platforms gain traction. One fruitful avenue of inquiry is to engage in close discourse with content creators who have achieved notable success on TikTok over an extended period. It would seem reasonable to suggest that such cultural workers have persistently and assiduously grappled with the logics of the platforms, amassed a portfolio of viral videos, and formulated well-defined notions about the logics of popularization in the context of platforms, shaped by their direct experience. Consequently, they possess a unique blend of specialized knowledge and experience that is typically beyond the reach of outsiders, including researchers.

For this reason, interviews were conducted with jazz musicians who have achieved notable success on TikTok. Subsequent to the corpus analysis, and in light of prior research on TikTok cultures, a number of specific inquiries emerged that were explored during the interviews. The following questions are of paramount importance to the research project: What are the key strategies for achieving and maintaining success in the realm of jazz on TikTok? How can jazz be effectively conveyed within the confines of a short-form video format? Which specific jazz repertoires and musical styles are particularly well-suited or ill-suited for presentation on TikTok? How should jazz musicians portray themselves on TikTok? Are there different requirements for female and male musicians, or for *white* creators and BIPoC? Additionally, it would be enlightening to understand the perspectives of musicians regarding the functioning of the platform. Do they deliberately attempt to interact with the platform's algorithms, and if so, in what manner? Does success also depend on audience reactions, in addition to the algorithmic moderation of the platform? To what extent can musicians accurately ascertain the elements that resonate with the TikTok audience? Do they remain in the realm of assumption or fantasy, or do they develop concrete formulas for success or optimization? What is the relationship between the musicians' work on the platform and their activities outside of TikTok? Does a platform-specific style of jazz emerge on TikTok? And could the representation of jazz on TikTok potentially alter the perception of jazz beyond the platform?

As the analysis of interview data is an interpretive process, with the results being contingent on the interpretive work of the researcher, it is imperative to ensure transparency regarding the specific research procedure. The ideal procedure would be seamless and intersubjectively comprehensible (Steinke 2017). To this end, the entire process – from the selection of interview partners to the analysis of interview data – will be described in the following.

The results of the corpus analysis serve as a basis for the selection of interviewees. In the course of this step, a number of musicians were identified who can be considered as being prototypical within the chosen field of investigation. It should be acknowledged that only a relatively small number of potential interviewees has been selected on the basis of this step in the research process. However, in the context of qualitative methods, this is not problematic or unintentional, as the aim is to engage with meaningful individual cases in order to be able to address the respective research questions at the required level of detail. The selection of these musicians can be justified by the theoretical sampling approach derived from grounded theory. This sampling method involves the deliberate selection of a few prototypical individual cases during the process of data collection and analysis. The specific cases are selected for analysis on the basis of findings that have already been generated in the research process (Strübing 2022, 594). This applies to the selected musicians, as their selection was based on the corpus analysis and thus empirically grounded.

In the initial stage of the interview study, an effort was made to establish communications with the musicians identified as prototypical in the preceding chapter: Ricky Rosen (@rickyrosen), Stacey Ryan (@staceyryanmusic), Laufey (@laufey), Rachel Chiu (@rachelchiu1), Ebony Loren (@ebonylorenmusic), Sam Ambers (@sam_ambers), Stella Cole (@stellakcole), and Erny Nunez (@young_crooner). It was not feasible to establish communications with the musicians via TikTok, as the platform's messaging functionality was limited to users who had established a direct connection. As the musicians maintained active Instagram accounts, an initial attempt was made to contact them through the platform's messaging feature, which yielded partial success. The musicians in question have amassed a considerable following on both TikTok and Instagram. Given the likelihood that they receive an overwhelming number of messages, particularly on Instagram, it was assumed that they would be unable to read or respond to them all. Consequently, efforts were made to contact them on other platforms where they had a smaller user base than on Instagram or TikTok. This is particularly the case with Facebook and X, but attempts to contact them via these platforms' messaging features were not successful. Some musicians provide email addresses on their TikTok profile or private website, and contacting them via email was successful in some cases. In other instances, multiple attempts were necessary to establish contact with musicians and arrange an interview, and sometimes involved different communication channels. As an alternative approach, efforts were made to liaise with the management or label in question if the musicians were already engaged in a comparable professional setting at the time of data collection for this study in the first months of 2022. This strategy proved effective in one instance.

However, only five of the musicians consented to be interviewed. Two of them disregarded all attempts to contact them, and one individual conveyed via their management that they were not inclined to participate in an interview. To conduct further interviews, additional musicians were contacted. The musicians in question were selected based on the results of the corpus analysis, which identified them as being among the top one hundred performers. Moreover, their videos display analogous characteristics to those of the musicians initially contacted, and they have attained a comparable level of success on TikTok: first, Brooklyn Stafford (@brooklyn.stafford), who appears on TikTok primarily as a pianist, but whose jazz-specific videos employ a similar repertoire to that of the aforementioned vocalists; secondly, trumpeter Kellin Hanas (@kellinhanas), who represents the humorous aspect of jazz on TikTok, while also performing older styles of jazz, particularly bebop; thirdly, singer Caity Gyorgy (@caitygyorgy), who is the sole interviewee not included in the top one hundred. The search for potential interviewees was expanded to include the hashtag #jazztok, which is frequently used by Caity Gyorgy. Her videos are similar in style to those of the aforementioned prototypical musicians and have a comparable reach. The following musicians were ultimately interviewed (in alphabetical order): Sam Ambers, Rachel Chiu, Stella Cole, Caity Gyorgy, Kellin

Hanas, Erny Nunez, Stacey Ryan, and Brooklyn Stafford. Some of the musicians have studied jazz at a post-secondary level, either at an undergraduate or graduate level (Caity Gyorgy, Kellin Hanas, and Stacey Ryan). One musician (Stella Cole) has pursued a degree in music and theater, also holding a B.A. in humanities. Another musician was pursuing a degree in music management at the time of the interview (Brooklyn Stafford), and one interviewee has a degree in French and Family Studies and works as an elementary school teacher (Rachel Chiu). Two of the interviewees were not engaged in a particular educational context at the time of the interview (Sam Ambers and Erny Nunez). With the exception of one individual, the interviewees had received some form of classical musical training from an early age. However, this did not necessarily extend to the field of jazz. The interviewees hail from the United States (Stella Cole, Kellin Hanas, Erny Nunez, and Brooklyn Stafford), Canada (Rachel Chiu, Caity Gyorgy, and Stacey Ryan), and England (Sam Ambers). With the exception of the Canadian musician Stacey Ryan, who currently resides in Los Angeles, they all reside in their country of origin.

Due to logistical constraints, primarily related to the considerable distances involved (the author was based in Graz, Austria at the time), the interviews were conducted digitally via Zoom and lasted between 60 and 120 minutes. One interview was considerably shorter, at only 30 minutes, due to the guidelines set out by the interviewee's management. The selected musicians have considerable expertise and knowledge in the context of jazz on TikTok, derived from their exceptional success in this field. They can therefore be considered experts in their field. Consequently, the procedure was based on the expert interview approach. The designation *expert* testifies to the musicians' distinctive role-related knowledge, which is shaped by their exceptional experience and unique position within the jazz music scene on TikTok (Przyborski and Wohlrab-Sahr 2021, 155–57). However, the term *expert knowledge* encompasses more than the mere possession of specialized knowledge by certain individuals. It also encompasses the potential for such knowledge to influence the practices of others. This implies that the actions of experts and their associated knowledge can also influence the actions of other actors (Bogner, Littig, and Menz 2014, 13). When applied to jazz on TikTok, this means that those creators who reach a particularly large number of people on the platform with jazz-related content can exert influence over the jazz-specific actions of other users due to the platform's emphasis on virality. To illustrate, the promotion of other users' viral videos or the incorporation of sounds and songs from the Great American Songbook may facilitate the dissemination of these musical works on the TikTok platform. It can be postulated that the selection of these particular pieces is based on the musicians' prior experiences and understanding of the strategies that contribute to success in this domain. The insights of successful musicians – or experts – therefore promise to elucidate the specific processes through which the image of jazz on TikTok was created and subsequently consolidated.

The structure of expert interviews is typically defined in advance and arranged in a specific order, with a set of topics provided to guide the discussion. However, to ensure a productive and engaging interview, it is essential to maintain a certain degree of flexibility and openness. It is recommended that narrative-generating questions be posed to the interviewee in order to facilitate the development of a natural flow of discourse and the potential emergence of topics that are already outlined in the semi-structured interview guide. Nevertheless, the interview can be moderated to a certain extent, with the occasional targeted follow-up question permitted, provided that a relevant topic is addressed by the interviewee in a cursory manner and not pursued any further (Przyborski and Wohlrab-Sahr 2021, 158–62).

The interviews were conducted in accordance with the aforementioned scheme and augmented by *conversational analysis* segments. This entailed watching TikTok videos – predominantly those created and uploaded by the interviewees themselves – along with the interviewees, with a view to furnishing tangible reference points for the description of their own activities on TikTok. This approach was predicated on the assumption that concrete examples would facilitate a more detailed discussion of the interviewees' own work, as well as the background and success stories associated with their videos. Moreover, the objective was to prompt the interviewees to recall the production of specific videos. The procedure was based on the video-stimulated recall interview methodology (Messmer 2015; Naujoks, Weiß and Händel 2021; Nguyen et al. 2013). In accordance with this methodology, the participants and the researchers watch videos that depict the participants in particular scenarios and subsequently speak about their behavior in these scenarios. Optimally, the participants would be able to recall the thoughts they had in these scenarios, at least in greater detail than they would be able to without the use of a concrete memory aid (Back and Klose 2018, 40). In the case of TikTok videos, the decision-making process underlying the production of the videos is already complete, or at least approaching completion. This implies that, for instance, a musical composition or the contextual setting can be selected in advance, whereas particular movements, visual elements, or auditory effects may be introduced extemporaneously. In the context of this study, it is of primary importance to facilitate a relaxed atmosphere through the use of media (TikTok videos), to provide concrete points of reference for the exchange of ideas about specific videos, and to create overall narrative stimuli.

The topics to be discussed in the interviews were developed and refined over the course of the research process. For example, the process of identifying relevant topics began with an initial exploration of the topic of jazz on TikTok. This was followed by a step-by-step elaboration through the reading of TikTok-related research literature, the development of initial hypotheses, and the corpus analysis. Consequently, the interviews were not entirely open, given the presence of pre-established research questions. However, a degree of openness was maintained, with the interviews structured according to the previously outlined criteria for guided expert interviews.

The interviews commenced with open questions pertaining to the musicians' musical careers, with a particular focus on their individual biographical connections to jazz. This was followed by an examination of the musicians' respective beginnings on TikTok, and a discussion of their first successes on the platform. Subsequently, inquiries were made regarding the genesis of their ideas and the production of their videos. This included questions about their sources of inspiration, the technical equipment they used, and the actual recording process. While these topics were discussed without viewing any videos, specific video examples were used in the following. For example, we first watched a video that was among the most popular content created by each interviewee. Subsequently, we analyzed the production process and the factors that could potentially have contributed to its virality. This approach facilitated the reconstruction of the specific mechanisms through which content gains popularity on TikTok. We then analyzed a video of the interviewees performing the jazz standard most popular on TikTok, according to the corpus analysis, namely "Fly Me to the Moon." The objective was to work with the musicians to examine the reasons behind the popularity of specific jazz repertoires on TikTok. Additionally, the discussion encompassed the potential benefits of performing one's own interpretations of jazz standards, the relatively low importance of original compositions and virtuosity, and the prevalence of humor in jazz performances on TikTok. Furthermore, videos were selected that illustrate specific techniques for fostering engagement, such as employing particular headlines or verbal announcements, or for enhancing the visibility of musicians through algorithmic practices, including the utilization of specific hashtags. To address the topic of body norms and associated cultural hegemony, a video by Stella Cole was presented for discussion with all the musicians (Cole 2020). In this video, Stella eschews the elegant attire and cosmetics that are hallmarks of her style on TikTok, prompting her to question the video's suitability for the TikTok platform. This video was discussed as an illustrative example of the influence of conventional beauty standards on TikTok. Subsequently, further inquiries were formulated in the absence of video footage. These queries concentrated on the role of TikTok in the professional lives of the musicians, particularly in contexts beyond the platform itself, and the extent to which TikTok differs from other platforms such as Instagram, on which the musicians are also active. The interviews were conducted in accordance with the aforementioned scheme, with the specific procedure varying depending on the individual situation. While it is possible that the interviewees were reluctant to divulge certain details, the interviews were notable for a remarkable degree of candor and willingness to share information. The fact that the musicians occasionally expressed identical opinions on specific subjects indicates that the interview data is highly reliable. Furthermore, the criterion of empirical content saturation was met, as it became evident during the analysis of the interview data that conducting additional interviews would not have yielded any fundamentally new insights (Strübing 2014, 32).

The interviews were transcribed, and as with the videos, the transcriptions were then analyzed using MAXQDA software. The utilization of qualitative data analysis (QDA) software enables the organization of coding and categories, and most notably, the retrieval of text passages and codes. This significantly facilitates the analytical process, particularly when dealing with large volumes of textual data. The objective of this phase was to construct a thematic category system that could be utilized to organize the topics that were derived from the interview data. The process of developing the category system commenced with a review of the first sentences from each interview transcript. As soon as a particular topic was identified in the transcripts, the corresponding text passage was marked to define a thematic category to which the text passage was assigned. This process was repeated, and the category system was continually revised and refined until every text passage could be assigned to a thematic category. Following the completion of this step, a period of several weeks was allowed to elapse before the interviews were reread in their entirety and the category system was revised once more. A number of changes were made, and the work on the category system was not completed until after this step (Hurst and Mayring 2005, 439). In essence, this was an inductive process, as the categories were derived from the empirical material and were not predetermined (Hurst and Mayring, 2005, 439). Nevertheless, various presuppositions or findings already generated during the corpus analysis guided not only the development of the interview guide but also the analysis of the interview data. Consequently, a strict model of qualitative data analysis was not adhered to, such as the method of qualitative content analysis (Hurst and Mayring 2005, 439). Rather, the procedure was operationalized throughout the research process in a manner that was sensitive to the specific subject under investigation. In the course of the analytical process, the following main categories were developed, which in turn were subdivided into numerous subcategories and sub-subcategories: (musical) biography; success on TikTok; production process; jazz repertoire on TikTok; specific demands on the musicians in the context of TikTok; TikTok-specific algorithmic practices; (gender) stereotypes, body norms and ideals of beauty; the relevance of TikTok for musicians and the extent to which TikTok could potentially influence the evolution of jazz or the perception of this musical culture – even beyond the platform.

As has already become evident, the selected interviewees constitute a relatively homogeneous group. This selection is not arbitrary but is based on previous empirical work. Furthermore, it can be justified by the specific research question of this study. This study focuses on the most popular musicians in the jazz field on TikTok. Consequently, it is essential to include these musicians in the discussion to gain a comprehensive understanding of the phenomenon under investigation. The empirical reconstruction of their knowledge should facilitate an understanding of the processes by which a particular image of jazz on TikTok is created and of the strategies that musicians must adopt in order to become visible on the platform with jazz-related content. Nevertheless, this methodology inevitably results in the exclusion of the voices of those who are not (or unable to be) as visible on TikTok. It is possible that the selected research design may contribute to the continued marginalization of specific groups on the platform. I am cognizant of this potential issue and wish to clarify that the findings of this study are limited to the specific area of investigation that I have chosen. It is imperative that the results be interpreted with a critical awareness of the cultural power of categories of social difference, such as gender and *race*. It is recommended that subsequent studies investigate other jazz-related phenomena on TikTok in greater detail. This should include musicians who, for various reasons, were not included in this study.

The results of the interview analysis are presented in the following chapter. Not all results can be presented in detail. The aim is to present the central topics on the basis of interview passages that are as representative as possible and to coherently link the content with each other.

6. Results of the Interviews

6.1 The Musicians' Careers

Which individuals are particularly adept at utilizing the TikTok platform to disseminate and engage with jazz-related content? Are they primarily jazz musicians with a classical education who pursued a conventional career trajectory from music school to conservatory? Or can musicians who have taken alternative paths also achieve success with jazz on TikTok? Given that the most successful jazz-related content on TikTok is a highly specific jazz repertoire closely aligned with popular music culture, the following section will also examine the role that genre concepts and the restriction to certain musical genres currently play for the interviewees and have played in their personal and professional development.

6.1.1 Early Musical Phase

All of the interviewees reported having been regularly surrounded by music from an early age. In some cases, they had already taken instrumental lessons at pre-school age, which is why their careers at first glance seem very typical of professional musicians. Some of these musicians were influenced by their parents and focused on classical music training, which entailed instrumental training based on the Western art music canon. Sam Ambers, for example, provides a detailed account of this: "I've played piano since I was five. [...] I've got all the Mozart, all the classical stuff on unlock and that's kind of my background. And I also played the drums since I was a kid as well." Rachel Chiu recalls a musical education that was markedly competitive, encompassing participation in various competitions with a particular emphasis on classical music:

> I started when I was five, my parents put me in piano lessons and I had a teacher that was pretty serious about piano. So I was very competitive. I would compete at music festivals. And I did the Royal Conservatory of Music. And I did classical piano the whole time until high school, until I graduated high school. (Rachel Chiu, interview by the author, 18 May 2022)

As Brooklyn Stafford observes, some of her classical training is evident in selected TikTok videos, particularly with regard to the technically demanding playing techniques:

> My classical training really helps me just get that dexterity down. I feel as a classical musician, you'll see a lot of classical influences probably in my videos because I started off very classical. I did, like, Suzuki, I've done Rachmaninoff and, like, those big jumps that I do, like, that's a more classical thing [...]. So yeah, my classical influences really helped me get the dexterity to [...] just jump around the piano. (Brooklyn Stafford, interview by the author, 28 June 2022)

For some interviewees, a strong connection to music was the result of the influence of relatives who were professionally engaged in the field. Kellin Hanas, for example, discusses her grandmother, "a choir director and music teacher," who introduced her to the trumpet. Caity Gyorgy recounts her academically accomplished and multifaceted secondary education, which encompassed a diverse range of artistic disciplines, including music. Despite an initial lack of discernible emphasis on jazz, her musical training ultimately constituted the majority of her academic pursuits:

> I went to the Calgary arts academy and so I was constantly doing artsy things in school there. And once I finished that school, I went to high school for grades 10 to 12 and I went to a performing and visual arts program at a public school in Calgary, where I was a part of three choirs. A concert choir, a chamber choir, and a jazz choir. I also took music theory and I did a lot of music there, like, ten and a half hours of rehearsal every week just for choir. Not to mention there was a quite a few hours of theory work as well to do. And lots of concerts to prepare for and so I was really, really involved in music there but it wasn't jazz. (Caity Gyorgy, interview by the author, 19 May 2022)

Nevertheless, a background in formal education is not a necessary condition for success in the context of jazz on TikTok. Erny Nunez notes that he did not

undergo any formal training and instead developed his skills by teaching himself: "Musically, I've never been trained. I've never had any special type of training. I wish I did. But musically, I guess I sort of adapt to what I hear. I've been told I have a great ear. And so I guess the more I listen to things, the more I pick up my stuff."

6.1.2 Educational Background

The educational pathways of the interviewees following their departure from formal education are diverse. However, it should be noted that a number of the interviewees had already commenced or completed a professional jazz education at a conservatory at the time the interviews were held. Caity Gyorgy recalls her very straightforward and institutionally bound professionalization path, which included studying at a conservatory after a school education that already had a strong musical focus: "I have been in music school, I just finished my master's degree, I went straight from high school to my bachelor's degree, straight from my bachelor's degree to my master's degree. I've been surrounded by musicians for the last decade, essentially." Stacey Ryan also talks about her time at the "jazz school," as does Kellin Hanas, who also regularly plays in select orchestras such as the National Youth Orchestra, which gives young musicians the opportunity to perform at Carnegie Hall, among other venues.

Despite receiving intensive musical training during their childhood and adolescence, not all of the interviewees pursued further studies in music. Rather, they regarded music as a hobby that they pursued alongside their academic studies, albeit sometimes with great dedication. Rachel Chiu is a case in point:

> I've been to school for the past five years at the University of British Columbia and I didn't study music. I didn't study jazz. I didn't study anything like that. I studied French literature as well as family studies. Because I want to become an elementary school teacher. And music has always been something I liked, so it's always been on the side. (Rachel Chiu, interview)

Stella Cole talks about her time at "Northwestern University outside of Chicago," where she "was studying theater, double majoring in theater and international relations, minored in German. And I got a certificate in musical theater. That's everything what I was studying in college." Sam Ambers discusses his educational trajectory, which involved a period of exploration that vacillated between musical and academic pursuits. His initial aspiration to pursue a career in music following graduation was initially impeded by the advent of the global pandemic. However, the subsequent virality of one of his TikTok videos proved to be a turning point in his career trajectory:

> In the UK from the ages of 16 to 18, you can do a thing called A-levels which prepare you for university. I did those and whilst I was doing those I kind of decided in myself that I wanted to pursue music. So when I was 16 and 17, I used to do loads of kind of jazz singing. I would sing at restaurants and kind of old peoples' homes and wherever, singing, you know, Frank Sinatra, Michael Bublé, all of that. And once I finished A-levels at 18, I was, like: You know what? I want to give music a shot. So I tried for a good six, seven, eight months, but then COVID came around and that kind of disrupted every plan that I had and really put, you know, put things to a halt. So at that point, I was, like, okay, it's probably a bit more safe if I go down an academic route and have that as something that I can fall back on. So I moved to Bristol [...] and my plan was to do economics at Bristol University and I was going to go there. But then the video that I posted on TikTok took off about two months beforehand, and I very, very quickly kind of managed to gain a bit of a following and I was, like, okay, let me give myself one more year on TikTok and we'll go from there. (Sam Ambers, interview by the author, 22 September 2022)

The preceding paragraphs illustrate that all interviewees inherently possess a robust and biographically influenced attitude towards music, if not exclusively towards jazz. However, this does not necessarily imply a desire to pursue music as a profession through formal education. In contrast with the conventional academic perception of jazz, the traditional classical route via a conservatory education is evidently not a prerequisite for remarkable success in digital spaces with jazz-related content.

6.1.3 How Did the Musicians Get into Jazz?

The question of how the interviewees were first drawn to jazz is a significant one. It is common for musicians to discuss the individuals who had a profound impact on their musical development. These individuals may be family members or teachers who provided inspiring music lessons. Erny Nunez, for instance, recounts the profound influence of his father, who introduced him to jazz:

> An inspirational person who got me into it was my father. He sadly passed away last year in August. But, you know, as I was growing up that's all he listened to. He taught me just that this type of music was sort of, like, what he called the best music. He always used to say that this is the best music, right? You got to bring it back. (Erny Nunez, interview by the author, 5 August 2022)

Sam Ambers recalls lengthy automobile journeys with his mother en route to their vacation destination in the northern reaches of Scotland. During these expeditions, he was first introduced to the music of Michael Bublé: "My mum loved Michael Bublé and he had just brought out his *Call Me Irresponsible* album. And I remember basically that being on repeat, you know, the entire way up and I'd say that's quite a soft introduction to jazz." In the case of Sam Ambers, a family member provided a "soft introduction," whereas for Caity Gyorgy, a romantic love affair was the decisive impetus for a jazz career: "I sort of got into jazz because I started dating a jazz drummer, which is really silly. But I actually really ended up liking the music." Other interviewees, particularly Rachel Chiu and Kellin Hanas, highlight the impact of their music instructors on their enthusiasm for jazz. Notably, Stella Cole offers an intriguing perspective, suggesting that her transition to jazz singing was largely influenced by her presence on TikTok:

> I wouldn't have called myself a jazz singer before the pandemic, actually. I've been singing this music for a really long time, but my focus was more on musical theater until I realized through doing TikToks and singing it more that this was something I really loved to do, just as much as acting and musical theater. So I started doing it. (Stella Cole, interview by the author, 1 August 2022)

This demonstrates that the experiences gained on the platform can significantly impact the future career trajectories of young musicians, including their stylistic preferences.

6.1.4 Musical Idols

In addition to significant figures in their personal and academic lives, the interviewees also identified numerous musical influences that played a pivotal role in their professional development. These include well-established singers who launched their careers several decades ago, such as Ella Fitzgerald, Frank Sinatra, and Judy Garland. Notably, Michael Bublé, who did not achieve prominence until the 2000s, is also frequently referenced. As the majority of interviewees were singers, it is unsurprising that singers are almost exclusively cited as idols. Looking at the results of the corpus analysis, however, it is striking that these are mainly musicians who can be assigned to mainstream pop music. In contrast, there is no mention of musicians who are more likely to be considered canonical in jazz-specific, specialized, or niche discourses. With regard to Judy Garland, Stella Cole even suggests that the singer may not be accurately described as a jazz musician:

> Judy Garland, she's not really a jazz singer, maybe some people consider her a jazz singer. I don't know. I never have, though. But I've been watching older musicals since I was two or three years old. Like, Mary Poppins in *Sound of Music* and *Meet Me in St. Louis, Wizard of Oz*. And then at Christmas time *White Christmas*. That was a lot of my exposure to the standards first, through watching old musicals and, you know, Judy Garland, I just grew up listening to her voice. (Stelle Cole, interview)

It is evident that an introduction to jazz in childhood and adolescence does not necessarily have to be achieved through exposure to jazz musicians who are presumed to be style-defining and pioneering in jazz circles – again, the "jazz icons" (Whyton 2010) mentioned in chapter 5.4 come to mind. The interviewees identified musicians who offer a "soft introduction to jazz," as Sam Ambers puts it, as being of particular importance. As evidenced by numerous interview passages, the interviewees perceive these artists as being primarily singers of a certain generation who achieved success around the 1930s, 1940s, and 1950s. Caity Gyorgy recalls her initial exposure to such vocalists through a compact disc released by Starbucks: "Starbucks used to make these CDs with compilations, and this one was great. And I sort of got introduced to a lot of different singers on that one, like, Blossom Dearie and Abby Lincoln was on that one and, of course, Ella [Fitzgerald] and Sarah [Vaughan], and Billie Holiday. It was really great. Actually, it was kind of a good opener." In this context, Sam Ambers identifies Michael Bublé as a particularly relevant example. His examination of Bublé's music also prompted him to examine the recordings on which Bublé's adaptations are based in greater detail: "Once you start listening to Michael Bublé, you start listening to the original versions of the songs that he covers and the alternate versions. And you get to your Frank Sinatras and then your Nat King Coles and your Chet Bakers. And it kind of goes from there."

6.1.5 Attitude to Musical Genres Other than Jazz

The interviewees demonstrate a broad and inclusive approach to musical taste and practice, engaging with a diverse array of genres beyond conventional forms of jazz. Some of the musicians interviewed underscored the centrality of jazz in their own musical endeavors, while simultaneously affirming the value of other musical forms. Caity Gyorgy offers a perspective on this phenomenon: "I think jazz is really where I found my calling. [...] I listen to lots of different kinds of music but jazz is what I sing." The majority of the interviewees also post songs from other genres on the platform, although the focus is on jazz. This is exemplified by Erny Nunez, who also performs country songs on TikTok: "I do like country. I like lots of different genres. But I want jazz to be my main type of genre." Brooklyn Stafford underscores her determination not to confine herself to a single musical genre. She also asserts that she does not necessarily identify as a jazz musician: "I wouldn't consider myself, like, the jazz musician to go to, but I consider myself just a musician as a whole. I play pop, I play jazz, I play different genres. I play Latin music, right? I consider myself a musician and not just limited to one genre."

This stylistic openness appears to be automatically reflected in the interviewees' TikTok profiles. It can be assumed that the diverse range of musical styles to which these musicians have been exposed throughout their lives has influenced their work on the platform. Sam Ambers offers insight into this phenomenon:

> It wasn't just jazz when I was growing up. I mean, it was a lot of Michael Jackson and Stevie Wonder, I guess more recently Bruno Mars. And yeah, I mean, it's been a whole variety of music. And that's kind of reflected in the TikTok stuff. You know, it's quite varied and it's not just jazz stuff. (Sam Ambers, interview)

Kellin Hanas posits that stylistic openness has become a necessity for jazz musicians, particularly in light of economic considerations. The live music market, which is of paramount importance for professional musicians, is increasingly demanding a diverse stylistic range from musicians:

> I've played funk gigs, more pop style stuff. I really am interested in musical theater and Broadway style, and I like to write songs in that style with, you know, lyrics and stuff. [...] Especially nowadays, it's important to be able to play everything. I want to be the most versatile player that I can be because then, you know, you get hired for everything. (Kellin Hanas, interview by the author, 23 June 2022)

Furthermore, Kellin maintains that it is imperative for the advancement and long-term sustainability of jazz to embrace influences beyond the conventional boundaries of the classical jazz canon: "I used to be such a jazz purist because I was, like, if you don't listen to old big band swing from the 40s, you're not real. [...] That stuff's great for what it was for the time and it's classic and we still listen to it. But the music has to move forward."

The interviewees all indicate a willingness to engage with musical forms beyond the domain of jazz. A majority of them express a desire to transcend the boundaries of jazz as a singular musical practice. These individuals represent a generation that did not necessarily grow up with a predisposition towards a single musical genre. Instead, they evince a proclivity towards incorporating diverse influences into their own musical endeavors. This reflects tendencies towards stylistic pluralization, which have been repeatedly described in studies of contemporary jazz discourses for several years (Burkhart 2019; Knauer 2018; Solis 2019) and which have always been a key factor in the development of jazz (DeVeaux 1991). Jazz on TikTok appears to be relatively open in terms of style, and success on the platform is not contingent on the classical training of the musician.

6.2 Starting a Career on TikTok

The progression of each TikToker's career was discussed in detail during the interviews. The following questions were posed: How does a social media career start and develop? What was the impetus to becoming seriously active on the platform? When did the user's success begin and what events was it triggered by? And to what extent can success on TikTok be planned?

6.2.1 Boredom and Creative Energy During the Pandemic

The meteoric rise of the TikTok platform in the first quarter of 2020 coincided with the first lockdowns in the wake of the COVID-19 pandemic. It can be reasonably inferred that this remarkable success was at least partially attributable to the sudden and widespread isolation experienced by the global population at the time. TikTok may have served as a source of diversion during this anomalous period, as well as a fortuitous opportunity for creative endeavors within the domestic sphere (Abidin 2020, 77).

Numerous individuals interviewed for this study did in fact assert that they began using TikTok during the lockdowns, primarily as a means of occupying their time. Caity Gyorgy is one such individual: "I think probably around last year was when I started getting into it. I can't even remember when I joined. I think it was definitely during the lockdowns and stuff, because I didn't really have much else to do." During this period, the social situation for a significant proportion of the younger population underwent a considerable transformation. The financial constraints associated with pursuing their studies led to a shift in students' ability to afford to remain in their place of education. In some cases, their sole viable option was to return to their parents' home, where employment prospects were limited. Stella Cole provides a detailed account of her circumstances during the pandemic:

> I started posting TikTok videos when I was in college, still at Northwestern University outside of Chicago, and the pandemic happened around my junior year of college. [...] So all of that sort of stopped. I took a quarter off college and I ended my lease in my apartment in Chicago, and I moved home for a year to live with my parents. And they live, like, kind of in the middle of the woods and there's nothing going on. (Stella Cole, interview)

Although the exceptional circumstances of the pandemic may have contributed to the success of TikTok, the musicians interviewed had already gained experience on various social media platforms in previous years. Rachel Chiu, for example, states that she "had a lot of time to sit at home, and that's when I was forced to find something to do." But at the same time, she states: "Since I had social media, maybe when I was 16, I would post singing videos of me, on my Instagram story, but it was just for my friends. And then when COVID started, I started posting on TikTok." In this respect, the pandemic seems to have prompted some creators to test the possibilities of a new platform and to transfer content that has been tried and tested on other platforms to TikTok.

6.2.2 Initial Negative Attitude Towards TikTok

However, not all interviewees recognized the potential of the platform from the outset. The majority of interviewees initially expressed skepticism about TikTok, emphasizing that they had to be persuaded by various individuals to become actively engaged on the platform. At its inception, TikTok was perceived as frivolous and juvenile, with little to no long-term viability. Interestingly, this assessment also applies to the musician who has since achieved her greatest success on the platform. Stacey Ryan reflects on her early experiences with TikTok: "I mean, at the beginning, I was one of those people who were, like: I don't want to get TikTok. I think it's stupid. It's just a trend. But then one of my friends convinced me to get it, and that was in November of 2019." Additionally, Kellin Hanas asserts that she initially had no personal connection to the content disseminated on TikTok and its predecessor platform, Musical.ly. Moreover, she expressed skepticism regarding the platform's potential for success:

> TikTok used to be this app called Musical.ly when I was in, I don't know, early high school or middle school. And everybody was posting silly stuff, it was a lot of dancing and a lot of lip synching and I was just not into that. And I remember being, like, TikTok is so stupid. I'm never going to download it, it's never going to take off. (Kellin Hanas, interview)

The content shared on the platform was designed to be lighthearted and youthful, which initially prompted some interviewees to question whether they should present themselves in this media context at all. Caity Gyorgy, for instance, recalls her initial skepticism about TikTok: "I was very reluctant to get on the app because my little brother was using it and I was, like, maybe I'm too old for this or people just gonna think I'm cringy for doing this." Brooklyn Stafford emphasizes her initial indifference to the platform, saying that at first, she only posted "silly stuff" on TikTok and did not get deeply involved, "because TikTok wasn't my medium."

6.2.3 Other Content Creators as a Source of Inspiration

Despite initial skepticism, at least some musicians appear to have developed a concrete plan for their own activities on TikTok relatively quickly. Stella Cole, for example, recalls her excitement when she discovered other content creators on the platform who were succeeding with jazz: "I saw other people on there, having success and I was, like, this is so cool. I saw Stacey Ryan, she was actually one of the first creators that I ever followed, and she had maybe, like, 150,000 followers at the time when I was just starting. And I was, like, this is so cool. She's singing all these songs I love and people like it." Consequently, as Stella herself states, she became increasingly driven to emulate the actions of creators like Stacey Ryan. In this case, it seems reasonable to posit that a specific plan was devised and implemented over a relatively short period of time, which subsequently informed her own actions on the platform.

6.2.4 The Role of Record Labels

In some instances, the impetus for launching a TikTok career originated from the musician's record label. It is probable that a considerable number of individuals engaged in professional music-related activities have identified the potential benefits that TikTok can offer for distributing content, marketing musicians, and, ultimately, achieving commercial success in the future. Caity Gyorgy states that it was her record label that encouraged her to pursue a career on TikTok: "My record label was actually the one, they said: Caity, you need to get on TikTok, you need to start posting TikToks. [...] And because my record label told me to, I did." However, this is an exception, as the majority of the interviewees were not signed to a label at the outset of their TikTok activity. Rather, they only became involved in such professional networks as a consequence of their initial successes on the platform.

6.2.5 Continuous Work on TikTok

Despite their initial skepticism, the interviewees, and in some cases their labels, appear to have rapidly recognized the potential benefits of maintaining a presence on the platform. Consequently, some of the musicians interviewed dedicated a significant amount of time and effort to developing their TikTok presence, continuously creating content to achieve and sustain a desired level of reach. Stella Cole points out that for a while she "worked hard and posted every day and made so much content and spent a lot of time on it," and that her joy at her first successes was all the greater. Sam Ambers asserts that he invested a considerable amount of effort into the platform, particularly during its nascent stages. Concurrently, he acknowledges his initial lack of familiarity with the intricacies of TikTok, which contributed to his initial sense of uncertainty:

> I remember at that time it was such a crossroads because it was such a leap of faith because TikTok was so unproven. But it was so strange even from the first video that I posted, ever since that video for a good three or four months, I was posting every single day, and I mean every day. And I remember I was on holiday with my girlfriend at the time and I had, like, no followers and there was no reason for me to do these videos. But I was, like, I just need to do this. It's hard to explain, but I need to do these videos. (Sam Ambers, interview)

The fact that he felt compelled to produce videos for reasons that are difficult to explain or rationalize is related to certain statements made by other interviewees. It would appear that when content creators achieve a degree of success, they feel an urgent need to repeat that success and continue to produce content accordingly. In this context, Brooklyn Stafford describes the strong urge to remain successful with her videos, or to become successful again when her popularity waned:

> You can't stop. You can't stop because the biggest thing is consistency. If you want to be successful in life, it's always about consistency, right? Putting it out there, putting your best out there, because the more you put out, the more you're going to receive. And you can't get discouraged if your numbers are low. You try something different the next time, a different approach on the TikTok medium, and then you see what comes back at you. (Brooklyn Stafford, interview)

6.3 Success on TikTok

As evidenced by the preceding paragraphs, the construction of a prosperous TikTok channel necessitates a considerable investment of time and effort. It appears that in order to achieve success on this platform, it is essential to develop tailored strategies for self-presentation and self-positioning, while consistently providing novel content. The interviews indicate that achieving success on TikTok is seldom an immediate process. The interviewees discuss initial periods of low engagement and acknowledge that success was not immediate. What are the specific mechanisms behind this phenomenon? Additionally, how do musicians navigate the fluctuations in their popularity on the platform, which can be significant?

6.3.1 Before and During the Early Days of Success

First and foremost, it is of paramount importance for the interviewees that their content is actually noticed on TikTok. Videos are uploaded with the express intention of being seen by as many people as possible, a fact that is explicitly stated in several interviews. Kellin Hanas, for example, emphasizes that "the goal of a TikTok is to get it out and get it seen," and Brooklyn Stafford states: "TikTok is all about trying to go for getting likes, getting lots of views, because views mean more followers and more followers means a bigger audience, which means that people are going to listen to your content."

Nevertheless, it is not uncommon for musicians to require a considerable period of time before they are able to garner the attention of a broader audience. Stella Cole, for example, remembers that her first video "got, like, zero views and then maybe, like, four views after a few days. Nobody cared at first," and Rachel Chiu reports that she started out with only "42 followers" on the platform. Initially, some of the interviewees attempted to incorporate nonmusical videos into their presentations, though these were ultimately not successful. Sam Ambers offers the following insight on this issue: "In the past I had tried with TikTok, but it was mainly, like, terrible comedy videos, like, absolutely shocking sketches. I remember I used to spend hours on these sketches, like, recording them and then editing them and they would get, like, 100 views and it was just pointless." Stacey Ryan expressed satisfaction with the number of views she achieved early on, which now appear insignificant:

> I started posting a couple of things here and there, but just for fun. [...] And then I remember posting that video and I think it got, like, 1,000 views, and that's the most I've ever got. And I had, like, eleven comments and I woke up to that and I was, like, oh my God, I went viral. I got 1,000 views. (Stacey Ryan, interview by the author, 20 July 2022)

Caity Gyorgy makes a similar comment about some of her videos that "did pretty well in the beginning stages, [...] getting 40,000 views," which she considers a noteworthy achievement at the time.

As was the case with the first videos posted by the interviewees, the reactions to their first successful uploads were, at times, enthusiastic. Erny Nunez recalls his first viral moment on TikTok:

> It was around my birthday. [...] I think it was one of the songs, "My Way" or something. [...] I went to sleep overnight, and I was just, like, man, what if these videos blow up? I woke up in the morning and I just started getting notification, notification, notification. I was, like, what's going on? Did my thing get hacked or something? [...] I started scrolling and scrolling through comments and comments, saying, wow, you have a beautiful voice. Wow, wow, wow. And I honestly couldn't believe it. I was in shock because I've never really received outside feedback from people. (Erny Nunez, interview)

Erny's descriptions are in reference to one of his videos, which achieved a considerable number of views within a relatively short period of time. It is noteworthy that he was able to attain this initial success with a song ("My Way") that was made popular in particular by a version sung by Frank Sinatra, a fact that appears to be consistent with the findings of the corpus analysis. Other interview passages illustrate that the selection of appropriate repertoires can be a pivotal factor for achieving success on TikTok. This is exemplified by Stella Cole's account of her initial achievements and the subsequent creation of her first viral videos on the platform:

> The first video that went a little bit viral was me singing musical theater. A Stephen Sondheim song, "Everybody Says Don't." That got, like, 30,000 views, which seemed so much at the time after getting, like, ten views on everything. And then I just kept singing. I kept posting more stuff and my first video that really blew up had, like, 800,000 views. [...] And at that point, I had maybe, like, 50,000 followers or something. But when I got into the 200,000s, that was because I started to sing more jazz standards. [...] People really love that. That's when it got really big. (Stella Cole, interview)

This quote shows that the most successful jazz musicians on the platform did not necessarily prioritize jazz as their initial focus. While Stella emphasizes elsewhere that she "wouldn't have called [herself] a jazz singer before the pandemic" (see chapter 6.1.3), her focus on specific jazz repertoires can be attributed to the observation that performing jazz standards appears to have a significant impact in the context of TikTok. That is to say, a specific repertoire, which, according to the corpus analysis, is representative of jazz on TikTok, can be a crucial factor for success.

6.3.2 Going Viral on TikTok

The musicians interviewed have a similar amount of experience with viral videos, as indicated by the fact that they have already uploaded several videos with view numbers comparable to those referenced by Stella in the preceding paragraph. While there is no universal criterion for determining when a video can be described as viral, the 800,000 views mentioned by Stella in the context of jazz – at least as suggested by the interviews – appears to be a sufficient magnitude for the video to be considered a major success. However, all of the interviewees were also able to achieve views in the seven-figure range with individual videos, some of them even in the eight-figure range.

This raises the question of how a video gains such a large number of views. All interviewees emphasized that the process is extremely rapid and that the number of views increases markedly within a short period of time. "When things start to go viral, [...] it starts off with a good number of views and then it just kept growing and growing and growing," explains Caity Gyorgy, while Kellin Hanas observes that after a specific threshold is reached, there is a five-figure increase in views every ten minutes. In recalling the circumstances surrounding the virality of her most successful TikTok video to date (Hanas 2021b), Kellin offers the following account:

> I posted it, I went to rehearsal and then I looked at my phone during rehearsal and within ten minutes it got 10,000 views. And I was, like, that's weird. And then I looked at my phone again when I came back and I didn't think it was going anywhere. I was, like, oh, it'll probably get 20,000 or 30,000, this is great. Yay! Then I came back from rehearsal. I was sitting in my bed and I forgot that I posted it. So I opened TikTok and then I saw that I had 99 plus notifications and the video was at a million views. (Kellin Hanas, interview)

These descriptions align with the observations of Sam Ambers, who was the sole interviewee to achieve notable success with the first video he posted under his current account, a rendition of the Michael Bublé song "Feeling Good" (Ambers 2021a). Sam recollects the process of achieving virality:

> I remember I woke up the next morning and I saw a couple of notifications on my phone, you know, saying that a couple of people commented. I then went on the video and I think it had 550 views and about 85 likes and a few comments. [...] And I thought nothing of it. I mean, I saw that it had, like, 550 views and it was quite a good like to view ratio, and I was quite obsessed with that. So I was, like, okay, that's quite cool. [...] However, throughout the day I was kind of with my family and we were on holiday at the time and I was just checking my phone every so often and there was, like, an extra 1,000 views and the next 1,000 views and then 10,000 views. And I was, like, wow, we're on 50,000 views now. This is crazy. This is actually crazy. And then I remember it was the evening time of the next day [...], at that point, it was going up. Crazy amounts. Like, every refresh was, like, another 1,000 views. (Sam Ambers, interview)

Statements such as these show that achieving virality is an unpredictable process, which is particularly surprising for musicians who are just beginning their careers on the platform. The process does not always end when a certain number of views has been reached. In some instances, it continues beyond the boundaries of the TikTok platform. For example, Brooklyn Stafford notes that one of her most successful videos rapidly gained popularity on other platforms, which she believes contributed to the rise in views on TikTok:

> Views kept going, like, by 10k, 20k, 50k, it went pretty fast. And I realized that it got even more viral when I started to get tagged on Instagram, on different musicians' pages that had a lot of followers. [...] That's when I knew it kind of went off. [...] It was during that time that I realized that it hit the 1 million views mark. (Brooklyn Stafford, interview)

While such bursts of popularity typically occur within a few days, Caity Gyorgy has observed that some videos appear on the For You pages of numerous users even months later. This phenomenon has the potential to result in a significant increase in views. In conclusion, it appears that the interviewees were not prepared for the viral distribution of their content, particularly at the outset of their TikTok careers. It also seems that there is little that can be done to substantially influence the process of going viral.

The impact of the sometimes considerable number of views on the number of followers remains unclear. Videos uploaded by content creators on the platform are not automatically visible to all followers of the corresponding profile. Consequently, a high number of followers is not necessarily indicative of a substantial reach (Zeng and Kaye 2022, 80). However, the musicians interviewed indicated that their number of followers had increased, albeit to varying degrees, as a result of the virality of some of their videos. Consequently, particularly successful videos can, at the very least, ensure that they are seen by as many people as possible in the long term, provided that the users decide to follow the respective profile. Kellin Hanas reports that the expansion of one's social media following can occur with remarkable swiftness: "I remember every day it was increasing, I'd just get 20,000 followers in a day. And I was literally, this is insane." Kellin's assertion indicates that this expansion is not a linear phenomenon, but rather a gradual process. Between April and September of 2022, the interval during which the interviews for this study were conducted, the two musicians with the highest number of followers and video likes were Rachel Chiu and Stacey Ryan. Both recall the surge in followers attributed to the proliferation of viral videos, citing the remarkable growth in the number of TikTok users who initiated following their profiles as a consequence of specific videos. Rachel Chiu states: "In the course of one year, I grew about 400,000, probably. It's very sudden, it's not constant, where all these people keep following you. It's every video, if it does really, really well, a lot of people follow me." Stacey Ryan cites her most popular video at the time of the interview as an illustrative example. This is a performance of her original composition, "Don't Text Me When You're Drunk," (Ryan 2021b) in which she invited other users to duet with her, resulting in the generation of tens of millions of views. The video ultimately facilitated her transition from the TikTok platform to a broader audience. Stacey delineates the process as follows:

> When you're just posting, it grows pretty steadily. Most of the time it's slower. But when you have a viral moment, it spikes so quickly and then it'll go back down and kind of just keeps growing. I feel, like, every viral moment I had gained me a lump sum of followers. Like, obviously the biggest being "Don't Text Me When You're Drunk," I think I gained 600,000 followers from that one moment. Because I was at 450,000 before I posted that and I reached, like, 1.2 million followers after that whole two-week long, three-week long process. And that blew my mind because, I mean, it's crazy. I was going up 100,000 followers in, like, a day. That's unheard of. (Stacey Ryan, interview)

In this context, Zeng and Kaye's reflections on a "virality-centric platform logic" (Zeng and Kaye 2022, 80) are validated. As evidenced by the statements of the interviewees, the sometimes considerable increase in the number of followers is primarily contingent upon discrete instances of viral dissemination. It would appear that TikTok users do not typically search for specific profiles and then like them. Rather, the objective for musicians is to have their videos appear on the For You pages of as many users as possible in order to achieve virality. Consequently, success on the platform is initially contingent upon algorithmic selection processes.

6.3.3 Surprised by Success

All of the interviewees emphasize that viral processes cannot be predicted with any amount of certainty. However, Rachel Chiu posits that, within the first hour of uploading, it is at least possible to estimate whether a video will generate a high number of views: "I can estimate that it will do well in the first hour. So if in the first hour I get 5,000 likes, then I know it's gonna do really well." These findings are based on a prolonged period of observation of the activities on the platform. The musicians who were interviewed expressed feelings of being overwhelmed by their initial success. They stated that they never anticipated reaching such a vast audience with their content. Sam Ambers, for example, notes that he "didn't really understand what TikTok was back in the day," so he "didn't expect for one second that it would get any amount of views or popularity or anything." In a similar account, Erny Nunez indicates that he initially did not take his activities on the platform very seriously at all and only began to do so after experiencing his first success:

> Honestly, I wasn't really thinking of it very seriously, I sort of thought of it as, like, a joke channel where I just, like, made random content. But then after I started getting views and I started seeing how people actually enjoyed jazz, I was just, like, you know what? Let me just take it more seriously and let's see where it goes. (Erny Nunez, interview)

The fact that success on TikTok can only be planned to a certain extent becomes clear, for example, when the interviewees recall viral moments that were completely unexpected. In some instances, their most successful uploads were videos that, from the interviewees' perspective, did not align with their actual focus on the platform. These interviewees also stated that they uploaded videos such as these without any ulterior motives. Rachel Chiu describes this situation with reference to her most successful video at the time of the interview (Chiu 2021b) as an example: "Another video I posted, I think got around maybe 15 million views. And that one was also very accidental, like, it was my friends at karaoke one night and my friend just decided to film it. I just posted it for fun because it was fun. And then it's the best video I've had, like, in terms of views."

Moreover, several interviews indicate that the most successful videos are often those that the musicians initially did not intend to post. The subsequent resounding success of these videos often takes the musicians themselves by surprise. This illustrates that the occurrence of viral moments can be contingent upon a series of fortuitous circumstances. "It wasn't planned," says Caity Gyorgy about her most popular video at the time of the interview (Gyorgy 2021b), "I didn't mean to put it on TikTok." Sam Ambers says that in one of his videos, in which he performs the song "Fly Me to the Moon," (Ambers 2021b) he was just "trying to gauge the levels with the music and the voice and how I'd look on the screen," and had no intention of actually posting the video. Interestingly, this is also true for one of the most successful jazz clips on the platform: Rachel Chiu's version of Duke Ellington's "It Don't Mean a Thing (If It Ain't Got that Swing)" (Chiu 2021a), which had more than eight million views at the time of the interview. Rachel recalls the circumstances under which the video was created. She positioned her smartphone on a stack of books and encountered difficulties in initiating the recording process. Her account of the phone's low battery level also provides an explanation for the video's description, "My phone was at 1%🥴," which initially appears to have no correlation with the content of the video.

> I remember that day, I just finished class and I stacked up a bunch of books and then it was about to fall, so I was, like, I need to record it and had 1% left. My cam charged my phone because it's at the bottom, right, but I need to prop it up. So, I was like: I have one percent. I need to get this one, right? And I did it. And actually, I wasn't even gonna post that. I was filming that for my friends on Snapchat. That's why the quality is not the best, but I find that funnier because it did well. (Rachel Chiu, interview)

That is to say, the production of the video was, in fact, an accidental occurrence. However, the selection of the song was also influenced by chance: "I had my real book for jazz standards. I had it pulled out and I was just, you know, playing around and then I was, like, oh, I forgot about the song 'It Don't Mean a Thing.' And then I was, like, maybe I should record it for fun."

6.3.4 What Does Success on TikTok Feel Like?

In what ways is this – at times remarkable – success experienced by the musicians interviewed? As the preceding sections demonstrate, TikTok's viral-centric platform logic means that young people can sometimes be seen by a large number of users on the platform within a very short time – sometimes reaching several million video views within a few days. At the outset of their TikTok careers, these were extraordinary circumstances that none of the interviewees had previously encountered to a comparable extent. Consequently, some of the interviewees continue to regard these occurrences as being somewhat surreal. Erny Nunez states: "I can't believe that there are exactly 100,000 plus people that are watching one person. I find that really crazy." It is evident that the ability to manage such sudden popularity and the considerable, albeit transient, visibility that successful content creators experience is a learning process.

A flourishing career on TikTok is often accompanied by a certain degree of visibility that extends beyond the confines of the platform. In this regard, numerous interviewees recounted instances where they were identified and approached in a public setting. This phenomenon is regarded as a source of inconvenience by many, as evidenced by Kellin Hanas' assertion: "People were coming up to me at school being, like, I keep seeing your TikToks, you're famous now and blah, blah, blah, blah, blah. [...] It was just very strange." Sam Ambers recalls analogous scenarios and reports that he now regards quotidian interactions with individuals he does not know differently, given that he constantly expects to be recognized:

> The first time that I was ever noticed in public, I had 50,000 followers at the time. And I remember I was just in a car park of a supermarket and a kid put his window down and he was, like, oh, are you the guy that has videos on TikTok? And this was the first time that it happened. And I was, like, yeah, yeah. Yeah, I guess. I guess. [...] And then things started growing so quickly. I remember I'd go on nights out with my friends and, you know, there'd be points where I'd go into a club and, you know, I'd be basically having people with the phones out all the time or coming up to me. It's extremely flattering. But now if I walk down the street and someone stares at me for a particularly long time, I can't help but assume that they might know me, but they prob-

> ably don't because they're just looking at me. So it is weird. (Sam Ambers, interview)

The success of TikTok has had a ripple effect extending beyond the platform itself. The interviewees' initial surprise at becoming "short-video celebrities" (Kaye, Zeng, and Wikström 2022, 97) may be attributed to their lack of awareness of the platform's potential reach and its impact on their daily lives. Despite the initial surprise, it is also crucial to exercise caution and refrain from allowing the sudden success to precipitate an inflated sense of self-importance, as Rachel Chiu puts it. Although popularity may have been an unexpected outcome, certain habitual effects set in over time. For example, as a successful TikTok creator, Rachel received words of appreciation and even declarations of love in digital spaces. The challenge is to differentiate between authentic and virtual existence. When queried about her emotional response to success on TikTok, Rachel responds:

> It definitely changes. The feelings change because you start to become a part of this small group of people who live a weird life, where online, a lot of people know who they are. And in real life, they're not as well-known. Because people recognize me when I start singing or when I start doing the mouth trumpet. But outside of that, I'm just a regular person. And it's definitely a strange feeling to deal with, and not many people that you can talk to about. I remember when I had 5,000 followers and I told myself if I ever have more than 10,000 to not be carried away with it, that is just a number. There's a weird contradiction and almost like a paradox that I have to deal with where I have to remind myself that, you know, all these people that keep complimenting me and stuff. It's just motivation to keep doing better. It's hard not to let it get to you. You know what I mean? If 10,000 people are telling you online, "Oh, I love you. You're amazing. You're the best," it gets to you, you know, and then it really builds your self-confidence. Then it makes you feel, like, you're better than some people. So it's really important not to let that get to you. But also, the other side is that you have to be appreciative of the comments. You have to be thankful and grateful for these people who spend time writing these comments to you, right? So it's that weird space that I'm in. It's definitely strange. (Rachel Chiu, interview)

This quote illustrates the potential for sudden success to elicit feelings of anxiety. These concerns extend to the possibility of repercussions for users' professional lives. Some of the interviewees express apprehension that a particular appearance on TikTok could have a detrimental effect on their professional trajectory, particularly in contexts outside the music industry. Sam Ambers expresses these concerns about his first viral video, which he acknowledges was recorded and uploaded somewhat hastily. This video shows Sam Ambers in a hotel room, where he suddenly emerges from a closet and starts singing Michael Bublé's version of the song "Feeling Good," only to disappear back into the closet shortly thereafter (Ambers 2021a). The video is notable for the intensity of his gaze, the brilliant lighting, and the singer's exaggerated facial expressions. In retrospect, Sam indicates that he initially considered removing the video when its viewership began to increase exponentially:

> My phone was going crazy. I had people messaging me that knew me saying: "Is this Sam person you? What's going on here?" And at that point, some people were commenting on the jaw and being, like, oh, this guy is on drugs. And at that time, I was really, really, really anxious about coming across, because at that time, I was, like, I'm going to university. I don't want a video of me looking like I'm doing drugs out there. So I remember I went for a shower and I was just, like, staring at nothing in the shower, like, contemplating my life, being, like, do I delete it? I don't want there to be anything about drugs, so I was really, really anxious about that. But yeah, the whole process of it getting views was really crazy. And all of a sudden, you feel very exposed. You know, what was your private video is now a public video and people are commenting on it and it's out there and I can't take it back because other people started reposting it, like, they're downloaded it and reposted it on their account. (Sam Ambers, interview)

The interviewees' presence in digital spaces is subjected to critical reflection for a number of reasons. This reflection also encompasses a critical examination of the mechanisms through which social media platforms, and in particular TikTok, have become popular and the underlying logic of their visibility. This is particularly evident in a statement by Stella Cole, in which she articulates her thoughts about her visibility on the platform:

> I think there's something weird about going viral all the time. I don't know if the human beings are really built to deal with the way that social media feels and how it feels to, like, one day be getting, like, 5,000 comments on something and then maybe a week later ten comments and nobody cares about what you're posting. It's just an interesting rollercoaster that in the past I've definitely struggled with. When you're getting less interactions with

> your videos, it makes you feel, like, you're less valid or important as an artist somehow, which isn't true. It's just an algorithm on an app, most of the time. Funnily enough, when I was blowing up, I kept checking on the video a little bit, but I tried to kind of stay away from it for a couple of days because it was very exciting, but it was sort of unbelievable. Very shocking. The first time a million people see your face, you're, like, that doesn't seem real at all, because I'm just sort of living my normal life and there are a bunch of people on the internet seeing me. Weird. (Stella Cole, interview)

6.3.5 When the Audience Cannot Be Reached

Stella's above statement suggests that one of the main difficulties for some content creators in using TikTok is their temporary lack of visibility on the platform. The algorithmic logic that governs the visibility of content on the For You page inherently places content creators in a position of relative dependency with respect to automated content moderation, while simultaneously constraining their ability to exert direct control over the extent of their own popularity. As previously indicated, the primary motivation for many of the interviewees to create new videos is the desire for greater visibility on the platform (see chapter 6.3.1). As Kellin Hanas notes, lack of success leads to a decline in motivation: "It's a very addictive cycle on TikTok. [...] And obviously when you're not getting as much interaction from an audience, you're not as compelled to make TikToks."

A number of interviewees expressed regret at the temporary absence of viral moments. One common source of frustration is the inability to reach all of their followers. It is not uncommon for new videos to receive only a few thousand views, despite having seven-figure follower counts. Consequently, the significance of follower numbers is occasionally called into question. Caity Gyorgy, who had amassed approximately 90,000 followers on TikTok at the time of the interview, states: "It seems very grandiose when you look at it, but some of my videos only get, like, 6,000 views on them sometimes. So how influential are those 90,000 followers?" The musicians are fully cognizant of the fact that this has to do with the underlying logic of the TikTok platform: "Unless it breaks through to the For You page sometimes, followers don't mean a thing on TikTok," says Brooklyn Stafford, and Stacey Ryan emphasizes that TikTok is "very touch and go with views and stuff" because of these mechanisms.

The interviewees appear to have accepted this reality to a certain degree. However, feelings of frustration arise when musicians perceive a possible underlying agenda behind TikTok's content moderation practices. Kellin Hanas, for instance, hypothesizes that TikTok may, on occasion, deliberately impede the success of specific content creators. Despite having generated multiple viral videos over an extended period, she suddenly found herself unable to replicate this success and failed to maintain her usual level of engagement with her followers:

> For a while I was getting millions of views over and over. I was, like, this is great. And it really gives you that high. It's amazing. And then all of a sudden, one day, TikTok just ends it. They're, like, no more for you, which doesn't make any sense. I have 155,000 followers right now or something. They do this thing where they kind of cut you off from the views and so you go, oh, well, I got 5 million views on my last video, but this one only got 30,000. What happened? [...] I mean, pretty much unless you're one of those very few TikTok creators that just blows up and keeps going and going and going and going, TikTok is just, like, alright, you had your month of fame, now we're done with you. So it kind of stinks. (Kellin Hanas, interview)

Stella Cole articulates a comparable perspective, yet she also delineates the specific circumstances of relying on platform algorithms to disseminate her musical compositions:

> It can be frustrating when you have, like, 100,000 followers, but because of the algorithm, only 10,000 people are seeing your stuff and you're, like, I know that all 100,000 of these people or all now 275,000 of these people chose to follow me and want to know when I have music coming out. But I can't reach them. I can't reach out to every person because this is an algorithm and this is an app. Only, you know, 10, 20% of my followers are going to see this video. (Stella Cole, interview)

6.4 Professionalizing on TikTok

The pursuit of success delineated in the preceding section also signifies that the interviewees began professionalizing their work on the platform at an early stage of their TikTok careers. This may involve purchasing specialized equipment and software for audio and video recording and editing, formulating customized staging techniques on the platform, and developing a meticulous approach to the creation of new videos. Some of the interviewees derive income from their music or are pursuing studies in music. For these individuals, it has become a matter of course that professional and long-term work in the music industry necessitates, among other things, effective self-presentation in digital spaces. Consequently, the standards for their own TikTok videos are frequently exceedingly high.

6.4.1 Producing Multiple Versions of Single Videos

Despite the common perception of TikTok as a platform for frivolous content and a lack of association with professionalism, the results of the interviews indicate that content creators often invest significant amounts of time and effort into their videos. In discussing the production process of the videos, the majority of interviewees recalled the frequently lengthy procedures that precede the final published videos. It is uncommon for musicians to complete a video recording in a single take. Typically, multiple versions of a video are produced, with the number of versions varying considerably. In some cases, up to fifty versions may be created. The interviewees emphasize their rapidly increasing quality standards, such as Rachel Chiu with regard to her viral video of the Ellington song "It Don't Mean a Thing (If It Ain't Got that Swing)" (Chiu 2021a). Although Rachel initially indicated that she was reluctant to post the video and that the quality could have been better, she needed numerous attempts before achieving a satisfactory outcome: "I think I've filmed it probably 20 times before I posted the one I liked." Sam Ambers articulates a comparable viewpoint regarding his inaugural viral video (Ambers 2021a). He was acutely aware of the possibility of achieving success with this video, which is why the video is founded upon a certain degree of quality:

> I could pretend and say that this video was completely just off the cuff, but we did about 15, 20 takes. [...] There was a little bit of thinking behind it and it would take a long time to do because I'd either blink or the singing would be bad or there'd be something off with the recording. And so I remember I was getting really frustrated with that particular video. It took a long time to do it. (Sam Ambers, interview)

It is noteworthy that these two videos rank among the most popular jazz videos on TikTok, and both are included in the top one hundred videos that form the basis of the corpus analysis for this study. Upon initial observation, the production appears relatively straightforward, as no discrete audio tracks were recorded and no specialized lighting, such as a ring light, was utilized in the respective rooms. Nevertheless, even ostensibly straightforward videos can conceal a considerable amount of effort and a particular vision of the final outcome. In some instances, a sense of aspiration has led to the development of a relatively sophisticated system of video production. Rachel Chiu delineates the processes she has developed over time to create her TikTok videos:

> I film on TikTok with three minutes because you could choose 15 seconds, 60 seconds and three minutes. I always do three minutes in case it goes over time, so it doesn't cut me off. And then afterwards, I will trim it down but usually I record it before looking at it, I just keep recording it. I record maybe 30 times, sometimes. Nowadays, I don't have enough time. So I usually just record it once or twice, but back then I would record it up to 30 times. And then I will review each video and think, okay, I didn't like this part. Let's see if I can fix it in the next part. And then I slowly delete the ones I don't like, that I know for sure I don't like. I usually reduce it down to three options. And then with those three, I really sit on it and then I pick the one that I really like. (Rachel Chiu, interview)

Rachel is not alone in this regard. Brooklyn Stafford, for instance, delineates her production process in a comparable fashion. In particular, her description reveals the rigorous standards she sets for her musical and instrumental performance:

> I take a lot of takes when I do these TikToks. Sometimes I could take 50 takes just to get the right one because you only see the good, the end result. But I make a lot of mistakes. When I do takes, I mess up here and there. I don't have the energy right. I don't feel confident in a video. [...] I make a lot of mistakes, you know, and you have to do those takes until you get it right. (Brooklyn Stafford, interview)

It is evident that this high standard extends beyond the domain of musical performance. Brooklyn's delineation of the production process highlights the role of non-musical elements in the long time it takes to record a TikTok video. According to Brooklyn, the first step is to come up with a basic musical idea, with this initial idea usually coming from "screwing around on the piano." Subsequently, a preliminary version is recorded on Snapchat, as this application provides the required storage capacity. Only then does the actual production process begin:

> And then the third step is I kind of wait a day. Usually, I do not record the day that I come up with the idea because first of all, I need to practice it over and over and over to make sure I know it. So it usually will take maybe a day, tops two days for me to actually record it. Rarely, I do it the day that I actually come up with that idea. It's very, very rare. And then the fourth step is getting ready for playing the piano, right? So I do my makeup, I do my hair, I get my outfits together, and then I go to the practice room. [...] I get my ring light and my tripod and then I start doing the takes. [...] It can take three hours for me to get a take that I want because at the same time, I'm practicing it. Sometimes I make mistakes and I have to retake it. And then every single time I have to keep that same energy, right? [...] I can be really frustrated with the take. You'd be like upset, but then I'll be, like, smiling and be happy in the video, right? And it takes a lot of – I would say acting, but also, like, make it kind of, like, a meditation in a sense, where you have to not be angry or upset or frustrated when it's, like, an hour or two and you're not getting the takes you want and you're, like, oh, screw it. I was so close to getting that perfect take. You have to just, like, get your energy back up and do the same take again. And you're doing that process for three hours and it can be kind of grueling sometimes. But you have to find what works. Every single take could be the take, right? And you have to make sure that the energy is there. (Brooklyn Stafford, interview)

The objective for Brooklyn is to establish a specific ambience or emotional tone in her videos, conveying the impression that the instrumental performance and production are effortless despite the demanding and occasionally stressful production standards. This is a specific challenge inherent to the production of TikTok videos, a sentiment that is echoed by numerous other musicians. Kellin Hanas, for instance, underscores the significance of precision in timing, particularly in the context of her comedic videos. According to Kellin, when a video contains a punchline, it is of particular importance to time the delivery of the punchline with precision while simultaneously conveying a sense of relaxation and humor. In her most successful video, titled "depressive episodes? no. j a z z," which had been viewed over five million times at the time of the interview, Kellin plays a fast bebop lick over a stock instrumental track of the jazz standard "Cherokee" before the video concludes with her emitting a scream (Hanas 2021b). Kellin delineates the production process of the nine-second clip as follows:

> I remember doing like 30 takes of that video. [...] Because it's such a timing thing. And you have to do it in the right way. [...] I just remember being, like, ok, where is the right time to come in on the backing track? Do I like what I played? Because, you know, every time it's going to be different and I'm going to be, like, well, I hated that. Let's try something else. You know, I was very self-conscious about the playing part and then also with the scream. If the scream is the comedic part of the video, then that has to be timed correctly and it has to sound right. And, you know, I would scream and I'd be, like, that was not the scream, that was too loud of a scream. That was too high pitched of a scream. You know, because you also don't want to annoy people. And I was very aware of that. I was, like, this has to be done in a specific way or people are going to get annoyed. But I don't want it to come off as annoying. I want it to come off as funny. [...] I just remember doing it over and over and over again and I felt so bad for the people around me. They were probably listening to me play over "Cherokee" and scream for twenty, thirty minutes. (Kellin Hanas, interview)

A similarly meticulous approach is evident in several of her humorous videos. In another highly successful video (Hanas 2021a), titled "bop dooba doo da bwiada women in jazz babyy 🤠🎺," Kellin is initially seen without her trumpet, addressing her TikTok audience with a spoken announcement: "Hi, my name is Kellin, I'm an eligible single lady looking for an eligible single man. I can cook, I can clean, and I can..." This is followed by a cut to Kellin performing a rapid bebop lick on the trumpet over a jazz instrumental track. The video offers an ironic commentary on certain gender stereotypes that persist in the discourse surrounding jazz. Kellin underscores the importance of timing, particularly in the context of such a combination of verbal and musical elements:

> It took me a little bit to record that. It had to have been at least an hour or an hour and a half, just because of the playing. Well, and even the acting at that point at the beginning, [...] there were a lot of subtle things that I was doing, just in the way that I was standing, my tone of voice, my inflection, my phrasing. I did a little wink at one point and I was just, like, how do I get all these things? That stuff is calculated because I want to make sure that the joke lands, you know? So I filmed that one in two parts. This one had to be tiny bit edited. I filmed the first part by itself and then once I was happy with that, then I got the backing track out and started to record the trumpet part. The trumpet part always takes the longest because I'm so self-conscious about my playing. (Kellin Hanas, interview)

Producing TikTok videos can thus be a time-consuming process, not just due to the high expectations placed on the final result. Several interviewees noted that they create a significantly greater number of videos than they ultimately choose to post. Rachel Chiu, for example, points out that she has "400 drafts right now that I haven't posted," and that she only records these videos for personal use. Nevertheless, there are instances when copious quantities of videos are produced with the objective of ensuring the continuity of professional-grade content. When producing new videos, musicians are also sometimes dependent on certain external conditions, such as lighting, which have a not insignificant impact on the production of professional videos. Caity Gyorgy describes her production routines:

> On a typical day [...] I will usually make quite a few TikToks. I rely on natural lighting in my apartment, so there's, like, between the hours of around, like, 12:00 p.m. to 3:00 p.m. when I have really nice light coming into my window. And I will create content for those few hours, and then put them in my drafts, and save them for later for the next week. (Caity Gyorgy, interview)

Concurrently, the production of multiple videos on a single day can be regarded as a strategy for balancing the demands of TikTok with the obligations of a career as a professional musician. Stella Cole emphasizes that TikTok represents merely one facet of her professional endeavors, which must be aligned with her other commitments:

> I'll usually film my content because of the way that an artist's schedule works or just anybody's schedule works. It's definitely difficult to film a new TikTok every day. So most of the time what I'll do when I'm posting really consistently is film a lot of TikToks in one batch and then sort of post throughout a week or so. (Stella Cole, interview)

While the aforementioned processes illustrate a degree of professionalization and formalization, some quotes indicate that spontaneity is also a significant factor in video production. For instance, Rachel Chiu states that her video, "My phone was at 1% 🥲" (Chiu 2021a), which was exceptionally successful, was essentially spontaneous. In this video, Rachel accompanies herself on the keyboard. After a brief interval, the keyboard is augmented by a drum pattern stored on the device. Following the AABA form of the song "It Don't Mean a Thing (If It Ain't Got that Swing)," Rachel performs on the *mouth trumpet*, which is to say, she imitates a trumpet with her voice. Although the selection of songs was spontaneous and the recording process was largely improvised, as previously described, the artist did not determine the precise musical interpretation of the song until the final stage of the production:

> [It was] a lot of experimentation. It definitely started very different. It was in a different key. It was in a different tempo. I sang different parts of the song to see which one I liked. And then I did up to posting the whole thing. And then I added mouth trumpet in the end. And the mouth trumpet parts are never written. It's always improvised. So you can see sometimes it's a little, like, I scoop notes because I'm not actually knowing where I'm going. I'm, like, oh, okay, I guess I'm going to this note. So that one was just improv. Which is really fun to me. So I ended up posting that one. I actually have another version that I loved a lot but I didn't like the trumpet part because I messed up on that one. So I ended up posting that one. (Rachel Chiu, interview)

Additionally, other interviewees assert that spontaneity is a crucial factor for creative ideation. "The ideas just come to me randomly," says Kellin Hanas, and Caity Gyorgy points out: "It's not really planned, it's pretty random, I guess." Consequently, although the production process may occasionally be protracted, spontaneous ideas and flashes of inspiration continue to exert a significant influence.

6.4.2 Recording Equipment

As indicated by the interviewees, the process of professionalizing production procedures on the TikTok platform entails acquiring a certain degree of technical proficiency in the utilization of software and hardware for audio and video recording and production. This occasionally gives rise to particular methods of utilizing the equipment that extend well beyond merely filming with a smartphone. In certain instances, the interviewees provide a comprehensive account of the actual processes involved in recording and editing. For instance, Stella Cole delineates how the production process has become increasingly intricate over time:

> It is a kind of a complicated process because I'm recording the video and audio with my phone, but also recording the audio linked up to my computer to GarageBand. And so then I have to edit them together and get the sound

> sync up to the video and then export that back to my phone and back to TikTok. So a lot of them might use a little bit more of a longer editing process, but I think it is worth it for the better sound quality. (Stella Cole, interview)

The use of audio software for post-production is a common practice among the interviewees, although not all of them adhere to a specific editing routine. Instead, they employ different recording methods on a case-by-case basis. Caity Gyorgy outlines her diverse approaches to this process:

> It really varies from video to video. Sometimes I use my iPhone. However, I have an iPhone 8 plus, and it's, like, quite a few years old. And now my microphone is starting to not work. So, I think I might get an upgrade to a better phone so I can make better TikToks and also just have more space. Because I keep running out of storage on my phone, but I use my iPhone camera and microphone. [...] I have a few microphones that I will use. Like, if I'm feeling particularly adventurous than I will hook up my audio interface. I've got one I got at the beginning of the pandemic, a Behringer. [...] And I have a really beautiful Neumann microphone that I absolutely love. [...] I've got a really nice handheld that I'll use for videos. [...] I have an Audio-Technica microphone that I've used before. [...] And I'll just use that through the interface and then I'll go into Final Cut and I'll edit that way. I do a lot of stuff in Final Cut. I've gotten pretty fast at Final Cut over the past couple years actually, which is a very good skill to have. (Caity Gyorgy, interview)

Erny Nunez states that he initially used a karaoke machine to create his videos, particularly during the nascent stages of his TikTok career, before transitioning to alternative audio recording techniques. He recalls that the recording process was particularly time-consuming during that period:

> When I first started, I didn't really have any equipment at all. And so I used to get this karaoke, like, speaker thing and I used to connect my other phone to it. [...] This took like 30 minutes to an hour every time to make a video back then. And I used to play the music recorded, like, over 20 times and find the perfect take. That's when I first started. As I got better and progressed, I got contacted by a fellow friend of mine [...]. He called me and said: "Hey, kid, I really want you to be successful. And so come down to my place." [...] I went down and he gave me this this whole entire, like, almost studio worth of equipment. [...] That was the first time I ever started using equipment. I was still so used to using karaoke machines because I'm not that good with technology. (Erny Nunez, interview)

Other interviewees emphasize that they deliberately keep the recording process as simple as possible. "I've never used a mic, I don't think, on TikTok," says Kellin Hanas, while Rachel Chiu asserts that her relatively straightforward recording process facilitates the production of new videos, which she perceives as a benefit:

> I think I got very lucky that I don't have to put in a lot of effort in my video process. And I think there are shortcomings to it but there's also advantages to it because it makes it more easy to do anyway, but I don't have a crazy technical process. I literally balance my phone on my piano. [...] And sometimes I have to stack up books and then just pop it up to whatever angle. (Rachel Chiu, interview)

6.4.3 Optimizing the Visual

While the aforementioned statements primarily address the audio recording process and its impact on video sound quality, the interviewees also highlight the significance of visual elements. Brooklyn Stafford, for instance, attributes her initial lack of success to the perceived lack of professionalism in the visual appearance of her videos: "A lot of them didn't get a lot of views in the beginning. I noticed my lighting setup was bad. I didn't clean up my look. It was a different look back then." While Brooklyn's primary focus here is on her appearance, she also provides detailed insights into the technical aspects of production, particularly in relation to lighting. She subsequently began devoting attention to the visual aspects of her videos, investing in equipment with the objective of enhancing the overall professional quality of her output:

> I purposefully changed it up. I got a tripod, I got an LED light, because I wanted to make sure my face was fully in cover. And if you'll notice from the very first TikTok that I did to that one that would go viral, I did, like, a few things that were different, right? The lighting was different. I made sure to clean up my look because whether we like it or not, TikTok and social media is focused on whether you look good. (Brooklyn Stafford, interview)

While Brooklyn discusses certain concepts of physical attractiveness that are pervasive and socially prevalent in the context of social media, other interviewees also note the importance of visual details that are not directly related to aspects of physical attractiveness. Kellin Hanas, for example, describes how the camera angle can be a significant factor in the production of TikTok videos:

> Probably the most time-consuming thing is figuring out the angle that I want it at. People kind of underestimate what a difference the video angle makes. And I think, you know, obviously I want to present myself in the best way. I also want to make sure that the framing is good. So it takes a lot of me if I'm filming in my dorm room, I'll take a music stand and put my phone up on a music stand and just move it around different places and see where is the most desirable place to film this. (Kellin Hanas, interview)

In this context, Stella Cole discusses her experience in the entertainment industry outside of TikTok. According to Stella, looking professional is a standard expectation in the acting and music industries, and she incorporates this into her work on TikTok, which contributes to the perceived level of professionalism in her videos:

> I think maybe they're more professional because I come from a background in acting and musical theater. And when you're auditioning and sending in a self-tape, professionalism and looking professional is very important. Like, the way that your video is set up in the theater and the TV film world is very important. And so I've just sort of applied that same mentality to TikTok, I think. I have, like, a ring light that I use and sometimes if I'm feeling more creative or have more time than I'll, you know, set up a couple of different lights and get some colors going. [...] Yeah, I care about how it looks. I think that matters. And I think it adds to the whole vibe and experience. (Stella Cole, interview)

6.4.4 Critical Perspectives on Older Videos

In light of the aforementioned professionalization processes and musicians' quality requirements, it is unsurprising that some of the interviewees retrospectively criticize their earlier videos. The TikTok platform facilitates specific learning processes, whereby early uploads may appear inadequate to the interviewees after a few months or a year or two. The reasons for this critical perspective vary and can be linked to the vocal or visual quality of older videos, for example. Referring to his first viral TikTok video (Ambers 2021a), Sam Ambers acknowledges: "Looking back now, I cringe massively. I think anyone would cringe, to be fair. But, you know, in those earlier days, I didn't care what

I posted. There was no real-life implication to what I posted. So, you know, I didn't understand really what TikTok was back in the day." In contrast, Stella Cole offers a critical assessment of her earlier work, particularly with regard to her vocal performance:

> Watching that video, my first response is, like, oh my God, I sound horrible. [...] I've grown so much as a singer in the last couple of years. So TikTok is definitely an interesting way to, like, keep track of your progress. And it's interesting to look back at videos from a few years ago and be, like, I'm phrasing that so weird, why am I singing that up-tempo like a ballad, you know? (Stella Cole, interview)

6.5 Interacting with Users and the Algorithm

The process of professionalizing one's output on TikTok entails the development of a nuanced understanding of the platform's unique requirements. This encompasses not only navigating the intricacies of the platform's algorithmic system but also engaging with users. The musicians thus generally have both algorithmic and human actors in mind.

6.5.1 Trying to Identify Patterns for Success

All of the musicians interviewed described their focused and long-term observations of the platform's logic. Over time, as the interviews illustrate, certain patterns are discovered that are believed to make success more likely and which are then continuously repeated in a similar way. Consequently, the creation of platform-specific content is frequently predicated on specific assumptions regarding the functional logic of TikTok. The musicians interviewed individually identified disparate potential patterns of success, thereby demonstrating the impossibility of establishing a universal recipe that can be applied as often as desired in the context of jazz.

Kellin Hanas describes her experience in identifying patterns of success based on her previously discussed video captioned "depressive episodes? no. j a z z" (Hanas 2021b):

> A girl playing the trumpet, that's the first thing that catches their eye. They go: this is weird, because there's not a lot of us. [...] Then they go: oh my God,

> she's playing really fast. That's impressive. And then they go: why did she just scream? Four million views. So I was, like, this is a recipe that I can use. [...] If you want to get attention and views and get more following, you've got to do stuff that's going to catch people's attention. So I would put on a really fast backing track and then kind of my theme became, like, I'm going crazy and I'm so stressed out and depressed. And people were, like, this is so funny. And also she plays the trumpet really well. So that's kind of what took off for me. And I did a series of videos of basically just play jazz instead of being depressed. And everybody liked that for some reason, and that was kind of my thing. [...] I mean, the goal of a TikTok is to get it out and get it seen. So I was just, you know what? What are the up-tempo tunes that I know that I can blow over and just do a short little lick and then do something afterwards, because that kind of seemed to be the recipe that was getting me the views. (Kellin Hanas, interview)

While Kellin explicitly refers to a "recipe," the other interviewees describe analogous scenarios in which they were able to discern specific formulas or patterns. Rachel Chiu, for instance, delineates this process through the use of the mouth trumpet, which is evident in numerous instances within her videos:

> I remember one time, a video did really well and it was me doing the mouth trumpet and I think I have a little bit of, like, the jazz pop style integrated into my mouth trumpet and so people really liked it and it was new and that was my very first video that blew up and it gave me the encouragement to try and do another one. (Rachel Chiu, interview)

Stacey Ryan asserts that she has identified a specific genre of video that consistently garners a significant number of views over time:

> I definitely saw all those videos, like, the looping harmony ones popping up all over TikTok. I love that idea because I love doing that, like, looping harmonies and stuff. But I was, like, I'm going to do it my way. I'm going to make it a little jazzier, a little more interesting. And the first one I did was "It's My Party." That was a big song on TikTok at that moment. So I did that in a way. I think it got, like, six million views or something, it did really well. I kept it pretty simple, but there was a little jazziness in it and people really liked it. So I did another one. [...] These are all getting a million plus views every time I posted one. It's a formula that works. So I kept doing it. (Stacey Ryan, interview)

6.5.2 "Find What Makes You Different"

One key assumption shared by several interviewees regarding potential recipes for success is the importance of unique selling points. They frequently argue that, in the context of TikTok, it is essential to differentiate oneself from the crowd in a distinctive manner while remaining aligned with the mainstream. The interviewees indicated that it is necessary to employ certain techniques to ensure that one's videos are easily recognizable. "That's a selling point. Find what makes you different and let that be the thing that stands out," says Brooklyn Stafford. Additionally, she emphasizes the importance of a "wow factor," which can be defined as a distinctive feature or quality that sets a video apart and attracts users' attention. In order to differentiate themselves from the vast quantity of content on TikTok, Brooklyn asserts that it is necessary to adopt a distinctive approach: "People are looking for a wow factor. People are looking for that wow moment. They need that wow moment. [...] If it doesn't have that wow moment where you take a little bit into something that you know and changing it to something better, people don't care how good it is, I would say." Consequently, it is crucial to avoid an exclusive focus on musical performance in TikTok videos. Instead, it is essential to consistently incorporate additional elements that can enhance the overall presentation, as Sam Ambers has emphasized: "That can be a look or a bit of humor or a bit of extra creativity or a bit of a niche to your videos. But you have to have something that you're kind of known for." This is precisely why, according to Kellin Hanas, even a technically demanding musical performance is typically insufficient to capture the attention of a significant number of TikTok users. With reference to her previously outlined recipe for success, which entails the integration of rapid trumpet techniques with a degree of humor, Kellin proceeds to outline her perspectives on potential avenues for the popularization of this approach:

> Most jazz musicians go on TikTok and they just play their instrument. But there are so many people who play the trumpet, it's just not impressive anymore. It's not really going to catch anybody's eye, like, you playing a jazz standard. People are going to scroll right past that. They're going to be, like, why do I care about this? [...] [Take] the first video that I had made with the trumpet in that context. [...] Nobody else had done something like that. There were plenty of people playing their musical instruments on TikTok. That's not new, but the combination of just something so incredibly ridiculous plus, you know, somebody playing the instrument at a high level, especially as a

> woman. […] There are so many factors in this video that are, you know, jump fact. (Kellin Hanas, interview)

As the processes of professionalization described above encompass enhancing one's visual appeal, it is unsurprising that certain methods of staging and performance style are perceived as potential determinants of success. Brooklyn Stafford, for instance, asserts that she has tailored her performance over time with the objective of rendering her videos more engaging to a broader audience:

> I also have changed the way I started performing, right? So I would start looking to the side. I'd be more expressive with my facial expressions, like, if I played something, I would move my body in that way. I would smile. I would put more energy into what I play because I noticed people like to watch people perform. They like to see the expressions on people's faces when they perform. And it takes you on the journey that I'm going through. So if I'm practicing, I'll probably just play practice like this: blank face. I won't make any faces, but when I get into performance mode, I purposely push myself more to make myself seem like a performer. And in a way it's like acting in a sense, too. (Brooklyn Stafford, interview)

The video features described by Brooklyn are evident in the majority of her successful videos.

Basically, it seems to be about being as creative as possible "within the confines of what TikTok thinks is popular. You have to be creative within what works on TikTok," says Sam Ambers, adding that it is especially important to "spice things up" and "add some kind of twist." It is not necessary for these techniques to be inherently musical in nature. To illustrate this point, Sam presents a specific example:

> After I posted my third or fourth video, it became a thing that I don't blink, I keep my eyes open. So that was an intentional thing to keep my eyes open. There's, like, a checklist of things that I do. First of all, I'd make sure that the song was okay, that the singing was okay, that the camera work was okay. But then I also make sure I wasn't blinking and that there were interesting points within the video that people can comment on. You know, they had to tick those boxes, the videos. So I could probably go through every video and kind of break it down and give you the reason behind it all. (Sam Ambers, interview)

This statement indicates that TikTok videos are occasionally meticulously crafted, based on specific criteria that have been established over time through a process of deliberate observation of the platform's logic. In this sense, TikTok has the potential to structure action in a way that is interpreted individually and creatively by successful musicians.

6.5.3 Attention Economy

The aforementioned techniques are of significant importance to musicians, regardless of whether they are musical, visual, or performative in nature. This is largely due to the fact that a fundamental assumption about how TikTok functions is that the logic of the For You page necessitates the immediate and widespread attention of users. It is presumed that videos must possess a distinctive and appealing quality to prevent users from swiping away from them immediately. Brooklyn Stafford articulates her assumptions as follows:

> We have to understand TikTok. When you're scrolling through, you have a very short amount of time to grab somebody's attention. People's attention spans are very short because of the way TikTok is. You scroll and if you don't like something within the first few seconds, you just skip, skip, skip. And in that way, it becomes so much harder to grab a listener's attention because you're competing with millions of other people, so many different pieces of content. (Brooklyn Stafford, interview)

As Sam Ambers notes, this phenomenon is clearly evident in the analytics data provided by TikTok to its content creators: "I can look at the analytics on my videos and the biggest dip of audience retention is within the first seconds. Most of the time it drops by a third because people just don't have the attention span. And if they don't immediately like a video, then you've got no chance of it doing well."

At this juncture, the capacity of the platform to facilitate the structuring of actions becomes evident once more, as content creators tailor their behavior to align with their interpretation of the platform's logic. One consequence of this is that some musicians opt to produce especially short videos, as their experience has shown that these elicit the most attention. For instance, Stella Cole posits that the platform logic of TikTok affects the duration of users' attention spans. Consequently, she has observed "that videos that are 30 seconds or a minute do much better for me." "It's got to be short and concise and get to the point. Otherwise, they're going to lose interest and scroll past," says Kellin Hanas. However, the production of new videos represents a delicate balancing act, as it is important to ensure that the content is not lost in videos that are too short. In this context, Brooklyn Stafford offers the following insight: "You have to keep it short. And you start thinking, like, how do you get enough content but not lose the attention of somebody when you're making, like, a jazz TikTok." In some cases, videos are created with the express purpose of garnering as many views as possible in a relatively short period of time. Kellin Hanas explains: "I'm always looking for a way to make them in a way that has that shock factor that will grab somebody's attention. And I always make them very, very short because that's how you get the most replays. You get the most watch time." In the absence of user intervention, videos on TikTok will automatically restart from the beginning, suggesting that this strategy is likely to be successful.

6.5.4 Algorithmic and User Imaginaries

The aforementioned assumptions and strategies for action are specifically applicable to the experiences of content creators as they observe the logic of the TikTok platform. These assumptions also pertain to the functioning of the platform's algorithmic system, which is responsible for moderating the content of all users' For You pages and thus exerts a significant influence on musicians' assumptions about potential pathways to success. In this context, concrete *algorithmic imaginaries* (Bucher 2018) of the interviewed content creators are revealed. Nevertheless, it is not always possible to distinguish between assumptions about the algorithms and assumptions about user behavior. Once a video has been included on a user's algorithmically moderated For You page, the subsequent action is for TikTok users to interact with the video. This may entail a range of actions, including swiping away from it immediately, leaving a comment, or duetting with the video. In this way, users can also contribute to the dissemination of a video. In this regard, algorithmic imaginaries or *algorithmic practices* (Abidin 2020) in the context of TikTok typically encompass both the platform's algorithms and the actions of its users.

It is of the utmost importance for musicians to gain an in-depth understanding of the inner workings of the TikTok algorithm. The algorithmic imaginaries that emerge during this process frequently influence their actions. The experiences of the content creators play an important role here. Assumptions about algorithmic logics are generally based on long-term observation of what happens on the platform and sometimes lead to differentiated strategies for action. In this context, Sam Ambers' statements are particularly instructive, as he indicates that he has engaged in extensive reflection on potential pathways to success. Sam primarily discusses specific and recurring patterns in his videos, which he hypothesizes are algorithmically prioritized. In this context, as Sam notes, it is crucial to exercise caution when integrating new elements into recently created videos, prioritizing the use of established techniques:

> When I've diversified and when I've tried to do different types of videos, they haven't necessarily done as well because, you know, with TikTok, you have to hit kind of, like, algorithmic baselines and you have to tick boxes. So, you know, if I'm trying something new out, then I think immediately that reduces my chances of something going viral because I'm kind of known and people I guess have liked videos where it's just the Samsung remote. So, the Samsung remote is the key indicator. I think it's important to have a distinctive feature about your videos in order to best kind of get the most out of the algorithm basically, you know. (Sam Ambers, interview)

The Samsung remote control is a prominent visual element in numerous videos created by Sam Ambers, including his first viral video (Ambers 2021a). It has become a distinctive and recognizable feature in his videos, often appearing in close-up with a voiceover from the cameraperson: "Samsung? Nah, mate. More, like, Sam sings." Nevertheless, it remains uncertain to what extent the repetitive presentation of this sequence affects the purely algorithmic prioritization, and whether the videos appear on numerous For You pages primarily due to such features. It is similarly conceivable that users who are already conversant with Sam Ambers are more prone to be captivated by such nuances, and that video intros are, in general, a source of fascination for a considerable number of users. It is equally plausible that these factors are operating in conjunction with one another. These processes are opaque even to those who have achieved success in content creation. However, Sam Ambers can at least report from his experience that videos lacking this feature tend to receive fewer views. Furthermore, as previously stated, Sam has devised a series of techniques purportedly designed to influence the algorithm and maintain user engagement:

> There are certain hashtags that do get pushed at certain times and your video will get artificial views, I guess, or views that it wouldn't necessarily otherwise get by utilizing a feature. Maybe it's a filter, a certain sound, a hashtag. You know, there are loads of little bits that you can really, really optimize, basically, to get as big an exposure as possible. [...] I kind of learned

> from what I did well, and again: it's about manipulating the algorithm, you know. Algorithm rewards you if you get comments, shares, favorites, viewer attention. And in order to do that you include little soundbites, you include little peculiarities, whatever. You include things that are just a bit interesting and a bit unexpected. And people comment on it and it keeps people engaged. (Sam Ambers, interview)

Similarly, Stella Cole underscores the significance of adhering to the logic of the platform. It is imperative to discern which content is most likely to be prioritized by the TikTok algorithm. This can be achieved by observing the strategies employed by other successful content creators. In the nascent stages of her TikTok career, Stella employed this strategy with particular efficacy: "When I first downloaded the app, I watched just all the singers that would come across my For You page and you just kind of see what works. It's not just about what people like. It's about what the algorithm likes, of course, and it's definitely difficult." In this regard, a deliberate examination of the algorithmic logic is a vital prerequisite for competing for visibility on TikTok. This is also a platform-related professionalization strategy. According to Sam Ambers, it is possible to gain a certain degree of expertise on how the TikTok algorithm functions:

> I spend a considerable amount of time thinking about TikTok, on TikTok, making videos for TikTok. I think, generally, I've got a decent idea of what will do okay. [...] This sounds like I'm showing off. I'm trying not to. But, you know, out of my last, I think, like, 30 videos, none of them have less than 100,000 views. So I feel, like, I've got a decent idea of what will work consistently well. [...] I won't post a video if I think it's going to do badly. Every time I post a video, there are check points that I want to hit. I want to make sure that it seems okay and if it doesn't feel right to me – there's a bit of an instinct as well that you kind of tend to develop. It's very intuitive, but it's like with anything, you know. I am now familiar with TikTok and the kind of landscape because I spent so much time on it. (Sam Ambers, interview)

6.5.5 Interacting with Users

The musicians who participated in this study all agreed that any kind of interaction on the platform can be beneficial in utilizing the algorithmic system and subsequently reaching a broader audience (cf. Cotter 2019, 903–04). "Any interaction on a video helps and it tells TikTok that people are interested in your video," says Stella Cole, which is why she sometimes posts promotional videos that simply ask users to interact. For example, she may request that users provide commentary on a video, share it, or engage in a duet. Stella states that she is skeptical about the value of such videos, but does acknowledge their role as a necessary component of her professional career:

> Sometimes you just got to be shameless about your self-promotion and make a bunch of videos that are short so that more people will see them and, you know, tell them you have new music out, and it never feels good to shamelessly self-promote and it embarrasses me sometimes. But it's just what you have to do in the world right now as an artist. So I'll do whatever it takes, you know, for that career I've always dreamed of. (Stella Cole, interview)

These assumptions about the necessity of constant interaction are, in turn, based on long-term observation of the platform's logic. With regard to this, Erny Nunez describes a learning process he has undergone in which identifying the optimal level of interaction is of paramount importance: "I found that commenting and replying actually sort of helps. I guess it boosts the video, you know, not replying to every single comment because I think that's kind of weird, but replying to some here and there, it does help."

6.5.6 Reading Comments

Observing the logic of the platform has resulted in the emergence of a number of platform-specific practices that are closely related to the various features of the TikTok user interface. Such analysis includes the regular and sometimes meticulous reading of comments written by other users about each posted video. The consensus is that this is particularly helpful in understanding what content might be particularly well received on the platform. "You look at the comments and you kind of know it works," says Brooklyn Stafford, and Stacey Ryan agrees that it's an effective way to better understand user preferences: "I definitely read the comments all the time. [...] And reading the comments does give you a good idea of what people like and what people want you to keep doing and what people don't like." Furthermore, an analysis of the comments allows for a deeper comprehension of the emotional responses elicited by specific songs or musical pieces among users. Stella Cole reports that she does not examine only the comments under her own videos; those under videos by other jazz musicians are also taken into account:

> People saying, like, oh, wow, this is so relaxing. Or a lot of people comment, like, oh, I'm, like, crying right now. I don't know why, but this is really moving me. And it's, like, it's just a video of me singing, like, "The Nearness of You" or something, you know? [...] All of those thoughts are coming from comments that I get on my own videos or comments that I see on videos of other people on jazztok that I follow. (Stella Cole, interview)

In addition, key assumptions about the *tricks* that several interviewees said were crucial for succeeding on TikTok can be traced back to reading the user comments. For instance, Rachel Chiu states that the comments made her realize that it is the mouth trumpet that elicits the greatest inspiration among many people. Regarding her viral version of "It Don't Mean a Thing (If It Ain't Got that Swing)" (Chiu 2021a), Rachel notes:

> It's a beautiful song, and I think people also recognize that, even if they never heard it, and never get the chance to, and then I was happy to sing it and incorporate my mouth trumpet which is always very popular. So I think that also really helped, if you look at the comments [...]. They [...] love, like, my little trumpet part and just the energy of it. (Rachel Chiu, interview)

Nevertheless, since the aforementioned tricks are not exclusively related to auditory details, content creators occasionally attempt to obtain information about non-musical nuances, which could also potentially contribute to an increase in popularity, by perusing the comments. In numerous videos, Sam Ambers demonstrates jaw movements that he later refers to as a key aspect of his style. After posting his first video, he states that reading the comments helped him to realize that this could be a recipe for success:

> I was still wanting to give myself a chance of the video doing well, you know, I tried to make sure that there were enough features within the video for people to comment about it. You can see that there's, like, some jaw movement and a lot of people were commenting about that, you know? So yeah, I tried to make sure that there were some kind of distinguishable features of the video and that it was a little bit shocking. (Sam Ambers, interview)

It would appear that Sam Ambers has been employing such strategies since the inception of his TikTok career. Over time, he has identified a number of criteria that he consistently incorporates when producing his videos. The comment feature can serve as a catalyst for the creation of new videos, as evidenced by the fact that musicians frequently engage with their own videos and with users' comments, thereby fostering further interaction. Kellin Hanas shares the following perspective on this issue: "You can respond to the comment and then do a video based off of that comment and whatever idea gave you. So it's just a good way to kind of connect your videos together and also engage the audience that was already there."

6.5.7 Duets

The duet feature is one of the most significant and frequently utilized aspects of TikTok. The capacity to use existing videos as a foundation for a duet enables the rapid dissemination of videos, with users who perform duets with popular videos potentially achieving a considerable reach. This feature, which fosters interaction and iteration, is emblematic of TikTok and has undoubtedly contributed to the platform's remarkable success. It is somewhat unexpected to discover that this particular feature does not occupy a key position in the perspectives of the musicians who were interviewed for this study. Only on rare occasions do the interviewees place a strong emphasis on the fact that the duet feature represents a highly significant element, enabling them to distribute their own content. Rachel Chiu offers the following perspective: "I think also what helps is that people duet it, and so people use my sound for their videos. And that also helps it grow a lot." Concurrently, Rachel indicates that she does not explicitly request that users engage in duets within the context of her videos. This is despite the fact that such interactions are a common practice among content creators:

> You know, there are people doing duets over duets. So I think it did work. There are also much more explicit ways to do it, which is by literally saying in the video: "duet me." I haven't done that yet. I don't know why. I just don't feel comfortable doing that. It just feels weird, but I definitely should try sometime. (Rachel Chiu, interview)

A review of the musician profiles reveals that duets do not occur frequently and are not a prominent feature in any of the profiles. Moreover, none of the interviewees commented in detail on this topic. At most, Sam Ambers, when queried, asserted that the production of duets is one of the platform-specific practices that could potentially be employed to attract the TikTok algorithm:

> [Duets are] massively important. I mean, from an algorithmic point of view, TikTok loves it when you engage with the features that it provides. So if you are stitching videos, if you are duetting videos, that's instantly a thumbs up from TikTok. Secondly, if you duet a video that is really, really popular, that gives yourself a really good chance of getting views because the For You page works in a way that [...] [i]f you've already liked five videos of people duetting a certain video, if I then duet that video as well, there's a chance that more people will see my videos. So again, it's about being smart and what to do and what not to do. [...] I think in general, engaging with TikTok and its features is very important and it exposes you to new audiences. (Sam Ambers)

6.5.8 Captions and Questions

There are a number of additional strategies that can be employed to enhance the visibility of videos, whether with the objective of fostering user engagement or of facilitating the algorithmic prioritization of one's own content. One method for fostering audience engagement is to incorporate inquiries into video descriptions or text layers situated directly within the video frame. One illustrative example is a video by Caity Gyorgy in which she scats a cappella to the chord progression of the jazz standard "Bye, Bye Blackbird" (Gyorgy 2021a). Scatting means that she strings together sequences of syllables in a manner that is devoid of any textual meaning. The text layers in the video frame provide insight into her rehearsal process, indicating that this is her preparation for performing with a big band. However, Caity does not mention the song "Bye, Bye Blackbird." In the video description, she writes the following: "Can you tell which song I'm soloing on? Bonus points if you can count how many times I breathed!" Upon request, Caity concedes that this is an effective method for fostering user engagement. Consequently, according to Caity, this could be a strategy for influencing the TikTok algorithm in a manner that is advantageous to her:

> I think posing a question caption is usually good. When I'm looking at TikToks, I don't typically read the captions very often. But previous to this video, I'd been doing a few other series where I would sing, I would scat a cappella, and I would say: Guess what song I'm soloing on, based on how well or not I'm making changes. And with this one, I noticed – it was actually my boyfriend who was, like, how many times did you breathe there? I didn't really hear you taking any breaths. And I'm, like, [...] maybe that'll be, like, a fun thing for people to listen for something. And I guess that I should actually proba-

> bly do that more often because it was very effective because it makes people watch to the end. And for those that do know the song, then they can be, like, oh well, this is this song. If I did a good job of outlining the harmony, but the breathing thing was kind of good. I should probably do that more, it definitely creates engagement and creates more views. And I'm sure all of that melts together to boost the algorithm. (Caity Gyorgy, interview)

Inserting lyrics into a video frame is a relatively common practice. "Putting the lyrics on the video works well," says Stella Cole. Additionally, she says that superimposed lyrics introduce a visual component that can enhance the visual appeal of a video. Concurrently, this methodology provides users with direct access to the songs played on TikTok, as Stella suggests:

> My guess is that when you put the lyrics, it gives people one something else to look at, so they won't scroll past. And then the other thing is that when you read the lyrics of the song – anytime I'm listening to music, I like to read the lyrics to the song. Either while I'm listening for the first time or before just so you get a better idea of what the song is about and you can process it in a visual way as well as an oral way. So, you know, for those two reasons I think it's good. (Stella Cole, interview)

Consequently, one principal objective may be to make the videos more accessible to other users and content creators. Rachel Chiu presents a similar viewpoint: "That's why also the lyrics work sometimes, because people might not know that song, so they want to search up the lyrics." In light of the considerable amount of content on TikTok, creators sometimes take their own assumptions about audience convenience into account when producing their videos. At the same time, Rachel states that on occasion she deliberately excludes lyrics in order to differentiate herself from other content creators. Nevertheless, in view of the fact that she has discovered that lyrics resonate with a diverse audience and enhance the accessibility of the videos, she has elected to incorporate them on occasion:

> Sometimes it's hard to hear the song that I'm singing, like, the words. [...] When there's words associated to what I'm singing, people are able to relate to it more because they're reading the poem as it's going. [...] Sometimes you can't really hear the words and sometimes people comment: "Oh, those are the lyrics! I never knew." And I think it just looks cleaner because it makes it look like a video where they can read the lyrics and they can also sing along

> to it if they want to. I don't know, I always just did it because other people did it. I noticed that other people did it. I used to not do it because I didn't want my friends to think that I was trying to be like the other TikTok singers. But I realized that at some point it was needed, it was necessary, and it just helped. (Rachel Chiu, interview)

Occasionally, the musicians open the videos with brief interstitial remarks. These typically introduce the musical piece or address particular user feedback, frequently responding to song requests. Erny Nunez suggests that this methodology may be regarded as a prospective formula for success on TikTok. Based on a period of prolonged observation, he suggests that his announcements, which he presents at the beginning of the majority of his TikTok videos, contribute to enhanced visibility, particularly given their capacity to imbue the videos with a personal touch: "All my videos were introductory and I found that for switching from those to just start singing out of nowhere, there is a very big difference in views and whatever else that comes with that. So, I guess the more personal the videos are, I guess the better they do."

Another potential approach for making videos more personalized is to respond to user requests. It is not uncommon for users to leave comments indicating their desire to hear specific songs played by content creators, and for those requests to be fulfilled. In such instances, the user comments are typically displayed within the video frame, thereby indicating that the musicians are responding to specific song requests. According to Kellin Hanas, this is a "typical way that you can [...] gain more traction."

While these assumptions and corresponding strategies for action are fundamentally derived from observations of user behavior, the interviewees also report on their efforts to obtain specific insights into the inner workings of the TikTok algorithm and to adjust their own behavior in accordance with these insights. Stella Cole posits that the role of a content creator necessitates a certain comprehension of algorithmic logic, which is difficult to develop concurrently with her primary occupation as a musician, primarily due to time constraints. Consequently, she relies on external sources of information, and a community has emerged on TikTok itself that is dedicated to collating data about the TikTok algorithm, specifically for the benefit of content creators:

> The TikTok algorithm is always changing, so I always am sort of reading about the most recent algorithm and what's working. [...] There's a community on TikTok also of people who just talk about influencer marketing and the al-

> gorithm and content strategies. And I watch a lot of those videos because I didn't train in marketing. I don't know what I'm doing there. (Stella Cole, interview)

Furthermore, the data that TikTok automatically provides about video views and likes can be utilized to make predictions about future success. For instance, Sam Ambers states that he routinely monitors the "like-view ratio," which can now accurately predict the potential success of a video. With regard to one of his most popular videos, in which he performs a cover version of the Earth, Wind & Fire song "September" (Ambers 2021c), Sam offers the following observation: "2.7 views to one like, and it maintained that for a long time." In such a case, it can be assumed that success is inevitable, as he further argues: "If you get a video that has three views to one like, then you know it's going to do well. So the fact that it was 2.7, I was, like, wow, this is crazy. Generally, as a rule of thumb, if you get below four views to one like then it's a pretty solid video." These statements reveal a meticulous observation of platform activity, which is sometimes conducted with great precision.

6.5.9 Hashtags

Some of the interviewees demonstrate a similar degree of precision in their use of hashtags. However, in this case, there was a certain amount of divergence in the approaches employed by the musicians interviewed. Some of them expressed difficulty in understanding the functionality of hashtags and in gauging their actual impact on the popularity of their videos. Nevertheless, all of the interviewees stated that they frequently utilize hashtags to categorize their videos. Nevertheless, this practice is often driven primarily by the observation that the majority of popular content creators employ this strategy, as Sam Ambers acknowledges: "I think hashtags – I mean, I still don't really know. I kind of do it because everybody else does it." Kellin Hans notes: "It's important to some aspect. I just don't know to what point it is. [...] I don't know how much that really matters or helps because there's plenty of viral videos that go viral without any hashtags. So I don't know what that's about, but I do it anyways just to see."

Despite the ambiguity surrounding the impact of hashtags, Caity Gyorgy indicates that her record label occasionally encourages her to utilize specific hashtags. However, this has not resulted in the formulation of any discernible hashtag-related strategies on her part:

> [Hashtags] used to be kind of important to me, but now I just usually do hashtag #jazztok. And then I'll do, like, #scatting or, like, #greatamericansongbook, that kind of stuff. I don't use them as much now, but my record label has insights into certain hashtags that are popular for each week and we'll get sent hashtags. So sometimes I'll use those if they apply to my video, but usually I just do hashtag #jazztok and I pray. (Caity Gyorgy, interview)

One evident strategy is to select hashtags that are as close as possible to the musical repertoire being performed. However, even this ostensibly straightforward decision entails processes of observation and learning, as certain effects that are anticipated to result from the use of specific hashtags may not manifest over time, as Stella Cole observes: "When I first started on TikTok, you wanted to use as many hashtags as possible and hashtag things like #foryoupage, #fyp, #viral. That stuff doesn't work anymore. You have to use niche hashtags and you should only use, like, five right now."

Nevertheless, some content creators have also developed highly specialized algorithmic practices when dealing with hashtags. This is exemplified by Rachel Chiu's detailed account of her extended hashtag experiments:

> The typical hashtag #fyp – For You page – this one doesn't work that well anymore, because everyone uses it. I learned to try to be specific, but also general. What that means is that I can't be so specific that no one else uses it, but I also need to be specific enough that I'm not one of billions of videos; I need to be one in a million. That makes sense. [...] I have to make sure I strategize what hashtags that I use: that [they] describe what I do but aren't so specific. For example, #mouthtrumpet. I'm able to use that because it's quite specific, but there's quite a few people doing it. And it's a good number, it's up to a few million who use it. So, that one, if you click onto the hashtag, you'll see a lot of me as well, because I'm one of, like, the people who do mouth trumpet the most. And there's also other hashtags that I always use, like #acoustic, if it's acoustic. And then, whatever instrument I'm playing. And then after that, it's a lot of different ones. It would be the artist's name of the cover that I'm doing. Sometimes I would put in if it's jazz, or whatever style it's in. And usually, I'll put around four or five hashtags at most. I never do more than that. [...] I do have a secret account that is also a singing account, but it's anonymous and that's my experiment page. That's where I try all these crazy hashtags to see what works. I learn a lot from that one. (Rachel Chiu, interview)

This segment of the interview with Rachel demonstrates how content creators frequently engage in meticulous observation of the activities and interactions that occur on TikTok, as well as their deliberate experimentation with the platform's functional logic. Nevertheless, it is important to note that the strategies employed by Rachel are, by their very nature, somewhat speculative. This is a fact that content creators are aware of and that can, on occasion, cause frustration.

6.5.10 "TikTok Is Such a Mystery"

The preceding paragraphs illustrate that the particular methods by which successful content creators navigate the platform logic of TikTok can vary considerably. It is therefore erroneous to assume that there are universally applicable algorithmic imaginaries or practices. Nevertheless, all interviewees share a common motivation to comprehend the functional logic of the platform in the most optimal manner. This unquestionably gives rise to comparable, though not entirely homogeneous, behaviors. Moreover, all of the musicians interviewed emphasized that the precise functioning of the algorithmic system of the TikTok platform remains opaque to a certain extent, even for experienced content creators who are occasionally aided by their record labels. The interviewees highlight the opacity of TikTok, noting that their strategies on the platform are inherently based on a degree of speculation: "You never know what's going to take off on TikTok. You just post and see what happens," says Brooklyn Stafford, while Caity Gyorgy states: "I can't say how it works. I have no idea. [...] I don't know why they choose specific videos. I don't know." Stacey Ryan observes that periods of largely unsuccessful outcomes often follow on the heels of viral moments, even as new TikTok videos are constantly being created: "I just kind of kept posting and there would be some little viral moments that would pop up through all the other videos, or then I would just post videos for months with nothing that ever goes, you know, goes really viral. But you never really know what to expect on TikTok." Stella Cole concludes: "TikTok is such a mystery even to creators like me who seem to have figured it out. But I still don't totally understand all of the ways it works."

Nevertheless, the prevailing view is that it is possible to formulate preliminary hypotheses based on accumulated experience regarding which content is more likely to result in success. Rachel Chiu states that: "You never know which video will do the best. I've done it enough to kind of have an idea answer, to be able to estimate, but sometimes it's always unpredictable, you really never know which one will do well." According to Stacey Ryan, it is possible to "to speculate. [...] But at the end of the day, you don't truly know." Stacey has previously attempted to leverage the TikTok algorithm to her benefit, employing a variety of strategies. However, she has yet to identify a consistently effective approach. This experience has led her to a disconcerting realization:

> I've tried over these past couple of years to try to post at a specific time or look at when my engagement is up compared to when it's not. And I've tried different hashtags. I've tried no hashtags. I feel like I tried everything. [...] Writing stuff like putting a title in it, not putting a title in it. I've tried writing the lyrics to my original songs on the screen when it's playing. I've tried not doing that, and I feel like I cannot get this clear idea of what works and what doesn't. [...] We take so much time trying to understand this platform and we never will. (Stacey Ryan, interview)

Notwithstanding these almost resigned-sounding statements, the explanations provided thus far indicate that all the musicians interviewed have developed their own repertoire of platform-specific optimization strategies. These strategies have enabled them to achieve success on TikTok, at least in numerous instances. The fundamental challenge is the inability to predict the platform's reaction to specific activities at any given point in time, while still developing and pursuing targeted strategies. As Sam Ambers aptly observed:

> I feel like TikTok is this new, exciting food that nobody really knows. Nobody really knows the ingredients. It's like KFCs herbs and spices. Nobody knows the exact combination of the eleven spices. No one knows how important it is to have viewer attention. How important a like or a comment or a favorite is or a hashtag is. But I think cumulatively having an awareness of the components that could make you successful is very important. But kind of the ratio of how important they all are, I'm not too sure. (Sam Ambers, interview)

6.6 "The Secret Sauce": Jazz Repertoire on TikTok

The preceding sections have demonstrated that popular content creators interact with the functional logic of the TikTok platform in a multitude of ways. They formulate specific assumptions regarding which types of behavior, under what circumstances can help attract the attention of both the platform's algorithms and its users. It has been demonstrated that, while there is no single, universally effective strategy for achieving success on TikTok, musicians can develop comparable approaches. It is evident that a relatively homogeneous picture of jazz on the platform emerges when the most successful musicians are considered. They interpret the TikTok platform in a basically similar way. It can thus be concluded that the platform's affordances elicit particular behaviors, that content creators respond to this to some extent, and that a substantial amount of comparable content is produced as a result.

It thus appears evident that the emphasis on a relatively homogeneous jazz repertoire on TikTok, as evidenced by the findings of the corpus analysis, can be attributed to the particular socio-technical dynamics between the platform and the content creators. These relationships were discussed at length with the musicians who were interviewed for this study. The popularity of relatively old jazz standards from the Great American Songbook on TikTok is a phenomenon that warrants further investigation. It is also important to understand why original compositions and virtuoso improvisations play a relatively minor role in the most popular jazz-related content posted on the platform. The experiences of musicians on TikTok suggest that certain repertoires are better suited to a short-form video context than others. This raises the question of whether certain repertoires are prioritized based on the platform's algorithmic moderation. Finally, it is essential to examine the role that the aforementioned ideas about TikTok's algorithms and users' practices play in the selection of repertoire and songs on the platform.

6.6.1 Trends on TikTok

A number of interviewees emphasized the importance of aligning the production of their videos with existing trends on TikTok. This entails adapting their content to align with songs or stylistic features that have already gained popularity on the platform. The precise manner in which such trends emerge remains uncertain. However, the interviewees maintain a close observation of TikTok with the objective of identifying trends at an early stage and adjusting their content production in a timely manner. "Generally, with TikTok, you have to be very, very, very responsive to trends," says Sam Ambers. Nevertheless, it is not sufficient to merely release videos of covers of popular or viral songs. As the interviewees indicated, musicians must be able to identify trends and infuse their own distinctive style into songs and videos to differentiate themselves from their peers. Otherwise, they run the risk of getting lost in the sheer volume of content. Brooklyn Stafford explains: "You go find a trend and then you add to it. [...] You have to catch the trend wave. If something is really popular, here's the secret sauce. If something is really popular and you think you can add something that makes it better: boom! You're set. That's all it is. Follow the trends and see what happens." Rachel Chiu also stresses the significance of discerning and interpreting trending songs in a manner that is distinct from the prevailing trends. She has shared a number of videos on TikTok, showcasing her proficiency in playing the guzheng, a traditional Chinese zither. In addition to the mouth trumpet, Rachel has also become known for playing the guzheng while singing lyrics in Chinese on TikTok. She occasionally performs well-known melodies and offers a reinterpretation of them through the use of the guzheng. To illustrate, she elaborates on her adaptation of the TikTok trend song "I Hate All Men but When He Loves Me" (Chiu 2021c): "This is actually a trend that was going on on TikTok. [...] So I kind of incorporated that but instead of doing that song I did a Chinese song that was very well-known and a lot of people resonated with it and I got 1.4 million plays on it." The selection of repertoire is frequently influenced by prevailing trends or the general popularity of specific songs, even in contexts beyond jazz. Sam Ambers: "Most of the videos that I post are related to either a trending song or a particularly famous song," and Rachel Chiu explains: "And a lot of the songs that I find that I want to do is based on what's trending." Once more, this methodology is predicated on the supposition that adaptations of songs that have already been incorporated into TikTok trends are more likely to be showcased on as many For You pages as possible. Consequently, users may be prompted to engage with the content or dissuaded from swiping due to its popularity. Sam Ambers notes: "If something is already popular, then when somebody sees you do a video that they've already seen was good, then they're more likely to stick around. Hence it will do well. So it all feeds into itself." This further illustrates the interconnection between musicians' perceptions of algorithmic processes and their understanding of user behavior.

It can be posited that specific musical criteria must be met for songs to become trends and circulate on platforms. For instance, Jost, Schmidt, and Neumann-Braun propose that songs that proliferate on YouTube as cover versions must be musically accessible to the widest possible audience. It is evident that this is particularly the case with regard to mainstream songs, which permit a considerable degree of expression without necessitating a high level of virtuosity or theatricality from the outset (Jost, Schmidt, and Neumann-Braun 2014, 347–48). This appears to hold true for TikTok, as the platform logic strongly encourages imitation, particularly in relation to the features duet, use this sound, and stitch. In this regard, musical compositions that exhibit specific iterative and low-threshold musical characteristics are more prone to dissemination within digital spaces.

Similarly, Stacey Ryan makes assumptions about one of her viral videos, in which she performs the jazz standard "Fly Me to the Moon" (Ryan 2021a). This is a duet video in which Stacey reacts to an existing video. In the original, an individual is observed seated in the driver's seat of a vehicle while another individual outside strikes the driver's window with a hammer, seemingly to assess the window's stability. Due to the regularity of the hammer blows, the video was widely referenced by musicians, who utilized the hammer sounds as the rhythmic foundation for their performances. Stacey recalls: "That video was going viral because everyone was playing songs over it because it had a good bpm. So people could pick any song, not every song, but many songs would work. So I was, like, that's cool, I feel, like, I could do something fun with that. And then I saw another guy who did 'Fly Me to the Moon'." This shows that the actual choice of songs is sometimes based on trends as well as on specific musical features.

It is also evident that people operating within the music industry can influence the process of selecting songs. A number of the individuals interviewed stated that they work with managers, which is a relatively common practice among TikTok content creators who have amassed a considerable following and viewership. Those employed in the music industry are confronted with the growing necessity of examining the data yielded by digital media platforms and adjusting their strategic decisions in accordance with this analysis (Baym et al. 2021; Maasø and Hagen 2020). It is therefore unsurprising that content creators on TikTok frequently receive recommendations with regard to prevailing trends and particular song selections. As Sam Ambers observes:

> I know, personally, you know, from a lot of friends, that they've really struggled with the fact that they're being told to post a TikTok every day, and they're being told to do trends and things that they would never normally do. But because it's TikTok and because it works on TikTok, they're being told by their record label or by their management that they have to do it as opposed to doing a video about the new song that they've written because people just don't have the attention span. (Sam Ambers, interview)

The specific assumptions about platform logics and algorithmic imaginaries thus have implications not only for the musicians themselves, but also for music industry actors who are confronted with the challenge of monetizing their observations and experiences on a given platform. However, none of the interviewees could confirm any such influences on their own TikTok activities.

6.6.2 TikTok Affords Certain Musical Repertoires

The assumption among the musicians interviewed for this study is that success cannot be achieved equally with all jazz repertoires. The interviewees indicated that they have to adapt to the rules of the platform to a certain extent and, on occasion, make compromises in their choice of songs. One of the consequences of this is that most of the musicians interviewed do not focus exclusively on jazz on TikTok, but rather demonstrate a stylistic range that encompasses a variety of musical genres. Sam Ambers commented more extensively on this topic during the interviews than any of the other musicians. According to him, an orientation towards so-called mainstream repertoires is a decisive factor for success on TikTok:

> Jazz obviously has a place on TikTok, but in order to give myself the best opportunity to develop a bit of an audience on TikTok, I've always thought the key word has been mainstream, and the key word has also been versatility. It's so important to be versatile, to try and to appeal to as many different audiences as possible, to try and kind of expose yourself to as many people as possible. And that is why you'll see that I do videos across entire genres, you know, an entire range of genres because I want to kind of reach as many people as possible. (Sam Ambers, interview)

As Sam also notes, there are numerous niches on TikTok where creators with highly specific content can achieve notable success and attain considerable reach. This is evident when one considers the vast number of users and uploaded videos on the platform. Additionally, jazz performances on TikTok can be regarded as a niche phenomenon, particularly in comparison to the most popular TikTok creators, who have followers in the tens of millions (Kaye, Zeng and Wikström 2022, 95–96). As Sam Ambers observes, creators on TikTok must decide whether to focus on a narrower niche or to pursue a more extensive reach through the selection of a particular repertoire:

> In order for something to do well, it has to have some kind of mainstream relevance or popularity. It's as simple as that, you know. Of course, TikTok actually does a really good job of finding niches. So if you're in a particular niche and you're a fan of that particular niche, then, you know, there's a

> happy marriage of a creator posting videos about that. And you as a viewer finding those videos. TikTok is really good at being able to recognize what people want. But generally, if you want the bigger views, you have to do more mainstream things. (Sam Ambers, interview)

Adherence to the so-called mainstream may necessitate a degree of compromise in artistic expression. Sam Ambers emphasizes that his TikTok profile is "not necessarily representative of [him] as a musician." He further asserts that it is crucial to occasionally set one's personal musical inclinations aside and turn one's attention toward those repertoires that, based on one's observations, appear to be the most promising on the platform:

> I think TikTok has its ways of streamlining what works, and sometimes you have to follow what works as opposed to what you actually do and kind of what is authentic to you. So there definitely is a bit of compromise. [...] I think in terms of musicality and artistry, [TikTok] really streamlines what people do. (Sam Ambers, interview)

The emphasis on trends and *TikTok-ready* repertoires has, at times, led to the perception that deviating from these norms may entail a certain degree of risk. As Sam notes, there is a possibility of experiencing a substantial decline in reach on the platform as a consequence of such deviation:

> The moment you kind of step outside of what is the stream and what is mainstream in terms of content, the moment you do something that's different, it's a big risk because, you know, you don't have the security of a trend, you know. It's just your own video. So yeah, I think it has limited artistry to an extent. And I think artists in general try to cater for TikTok. (Sam Ambers, interview)

Sam Ambers posits that it is imperative for musicians on TikTok to concentrate on specific mainstream repertoires. The question thus arises as to which repertoire this refers to in the context of jazz. The results of the corpus analysis indicate that the most popular mainstream repertoire in this context is made up of the Great American Songbook and swing era compositions. These include songs such as "Fly Me to the Moon" and "It Don't Mean a Thing (If It Ain't Got that Swing)," which are associated with renowned figures such as Frank Sinatra and Duke Ellington. The popularity of established jazz standards on TikTok can be attributed to the musicians' perceptions of which types of repertoires will resonate with users and the reasons behind this resonance. In her argument, Brooklyn Stafford asserts that it is of the utmost importance to provide TikTok users with a sense of familiarity, while simultaneously imbuing the songs with a distinctive personal touch:

> The key to getting something successful on TikTok is to get something that a lot of people know and adding a different element to it. And I would say it has less to do with how good you are technically, right. You need to tap into people's emotions. You need to tap into people's familiarity. People look for familiarity, right, in music and pieces and whatever medium that you decide to play. It's all about tapping into that familiarity. And that's the secret sauce. It never failed once. [...] I'd say for jazz musicians trying to get more attention, more notoriety, it really has to come down to – can you put a spin on something that people already know? That's my biggest suggestion. [...] Being a jazz musician on TikTok requires you to keep pushing. And keep doing a different approach to things that people already like. (Brooklyn Stafford, interview)

As with the musicians' remarks on TikTok trends (see above), there is an emphasis on the need for differentiation and to create videos with recognition value. One particularly effective strategy – or, in Brooklyn's words, "the secret sauce" – seems to be to reinterpret relatively old jazz standards that are perceived as having a high degree of recognition among TikTok users in a highly creative manner. As Brooklyn Stafford goes on to explain, this is tantamount to a recipe for success. She exemplifies this with the song "Fly Me to the Moon," of which she – like nearly every popular jazz musician on TikTok – has uploaded her own highly popular version on TikTok (Stafford 2021a):

> "Fly Me to the Moon," right? That's more of my jazz kind of more centered songs, the "Fly Me to the Moon" piece that I did. And then again, that's also a very familiar piece. Everybody knows "Fly Me to the Moon," right? And once you add a different element to it, people love that. And I could do the same approach to a different piece. (Brooklyn Stafford, interview)

The other interviewees corroborate Brooklyn's assertion that TikTok users respond favorably to videos on the basis of their familiarity with specific songs. Stella Cole also emphasizes that the selection of her repertoire is significantly shaped by past audience responses. This is precisely why her musical output on TikTok occasionally deviates from what she produces outside of the platform: "In real life, I like singing standards that people don't know as well. But on TikTok a lot of people are interested in hearing songs that they already know." Caity Gyorgy also assumes that user familiarity with specific musical compositions is a pivotal determinant of success on TikTok: "People really like what's familiar, I think. At least that's what I've noticed over time. [...] People like the videos of the songs that they already know."

Stella Cole also hypothesizes that not all jazz repertoire is equally suited for dissemination on TikTok. The mainstream appeal of songs like "Fly Me to the Moon" is a significant factor, and many users associate nostalgic sentiments with such songs, as evidenced by the comments on her videos. It is evident that nostalgia plays a role on a relatively new platform like TikTok, which has been demonstrated to be used primarily by comparatively young individuals (Zeng and Kaye 2022, 80):

> Like I said earlier, people really like to hear what they know. Especially, I mean, a lot of people comment on my videos and tell me when I'm on my live streams and stuff that the music I sing brings them a lot of comfort and makes them feel a lot of nostalgia. And so I think that's part of the draw of the old jazz. Like, everybody knows the song "Fly Me to the Moon," whereas there maybe aren't as many people who would know a modern free jazz tune or really understand as much what that's about on TikTok, you know. [...] The jazz right now is not pop music, obviously, but you know, in the thirties and forties, fifties and even in the sixties, I mean, these standards were pop music. (Stella Cole, interview)

Nevertheless, there appear to be various manifestations of nostalgia. As indicated by the interviewees, nostalgic sentiments are not exclusively associated with older songs. They also encompass songs that, for the predominantly younger TikTok users, are linked to memories of their childhood and adolescence. One illustrative example is the animated series *The Fairly OddParents*, which first aired in the United States in the early 2000s. The show's theme song is performed by Brooklyn Stafford in one of her most popular TikTok videos. The video is a mash-up with the song "Friend Like Me," which was released on the soundtrack of the Disney movie *Aladdin* (1992). The caption to the video on TikTok reads, "The mashup you didn't ask for but needed 🥴🎶" (Stafford 2021b). Brooklyn offers an explanation for the video's success, stating:

> People really gravitate towards stuff that has a nostalgia factor for them. What I found to be the most popular things that really catch people's attention – sometimes it's not even jazz. But say, for instance, I did, like, a *Fairly OddParents*, that was my most popular video. It was my *Fairly OddParents* plus my – the Latin song remix, "Friend Like Me." And I think one of the reasons why it got so popular is because everybody really knows that childhood *Fairly OddParents* song. Everybody has that nostalgia of that, and once you take something that was so important to them and you do a totally different twist on it, it blows people's minds away, right? (Brooklyn Stafford, interview)

Once again, Brooklyn reiterates the necessity of infusing popular songs with a unique twist. This assertion is further substantiated by her reiteration of the core assumption held by content creators regarding the characteristics of successful videos. As Sam Ambers notes, the act of *ticking boxes* is a fundamental aspect of the interviewees' interpretations of TikTok. According to Brooklyn, there are three key elements at play: the *wow factor*, the degree of familiarity with the songs, and the potential of certain songs to evoke nostalgic feelings in users. "So there's three things, right? You have to have a wow factor. You have to have something that's already known, right? Like, something that somebody can attach to. If they know the piece, if it brings in nostalgia for them, if it has an emotional factor, that's very important."

However, the nostalgic sentiments musicians evoke in their audiences are not solely a reflection of their personal experiences with music. They can also be attributed to specific individuals, many of whom are considerably older than most TikTok users. As illustrated by Stella Cole, a considerable number of her admirers associate the Great American Songbook compositions with their grandparents:

> Even if it's the younger people on TikTok who see me and are, like, oh my gosh, my grandparents listen to this music, or my parents always listened to this growing up, or this was their wedding song. You know, they're American standards for a reason. They're standard songs that everybody knows, a lot of people know. And they're just the classics. (Stella Cole, interview)

This type of nostalgia, which primarily relates to individuals with whom the user has a close relationship and not necessarily to the user's own musical or media socialization, is likely to contribute to the popularity of jazz repertoires on TikTok that, at first glance, do not seem to have a strong connection to the lives of most TikTok users. Using "Fly Me to the Moon" as an example, Caity Gyorgy speculates that the piece is likely to evoke nostalgic sentiments in many individuals due to its association with memories of loved ones, particularly within the familial context. Caity herself is no exception to this phenomenon:

> "Fly Me to the Moon" [...] that one's just so popular. Everybody knows that song. Especially lots of older people here. Anybody who's, like, you know, above the age of, like, 60 knows that song and probably showed it to their kids, because that's the music they grew up with, that kind of stuff. [...] So I think there's sort of a comfort in that, I think at least for me, you know, like, I would listen to certain songs with my grandparents growing up and [...] this music makes me think of my grandpa a lot. Especially Frank Sinatra, my grandpa loved Frank Sinatra and the Rat Pack. I think maybe other people can relate to that as well. (Caity Gyorgy, interview)

Moreover, a number of jazz standards have achieved a certain degree of familiarity among younger audiences as a result of their inclusion in the curricula of educational institutions or private music lessons. Rachel Chiu, for instance, notes that a piece such as "Autumn Leaves," which she performs in one of her popular TikTok videos (Chiu 2022), has become a recognizable part of the musical landscape for many young people "because a lot of kids in high school had to learn, you know, jazz standards. So it's, like, almost me trying to bring that nostalgia back for them. Something that they haven't heard in a long time."

However, Stacey Ryan is quick to point out that the demographic of users who listen to jazz on TikTok is not limited to people under the age of twenty-five. In one of her most popular videos, Stacey Ryan performs "Boogie Woogie Bugle Boy," originally popularized by the vocal group The Andrews Sisters in 1941 (Ryan 2020). According to Stacey, individuals outside of the typical teenage and young adult demographic have commented positively on the video. She attributes this primarily to the fact that the video was posted on the For You pages of older users:

> I remember noticing that it was a lot of older people were seeing it. And it makes sense because that's a song from an older generation. [...] I remember just getting all these comments and you could just tell it was older people writing them. [...] I found it interesting that that video sought out that demographic because it's a really appropriate demographic, because it was showing it to people who actually heard that song when they were, like, my age or ten years older or 20 years older. So I feel like TikTok really, really did well with that one. (Stacey Ryan, interview)

Additionally, some musicians have expressed reservations about the nostalgic emphasis on traditional jazz standards. This is primarily attributable to the perception among some of the interviewees that there is an unreflective glorification of the past, which they believe may be a factor in the popularity of songs from the Great American Songbook on TikTok. Kellin Hanas argues that these songs do not play a role in the lives of many TikTok users. However, their popularity is concomitant with what she perceives as a problematic romanticization of life in the first half of the twentieth century:

> All of a sudden, because it's popular, people are romanticizing the 1940s, the 1930s, the 1950s. It's the same thing of, you know, you'll see these videos on TikTok of these girls being, like, here's my 1940s style, get ready with me and they'll dress up and do their makeup and then all the comments will be these young girls being, like, oh, I wish that I could live in the 40s. No, you don't. You're a woman. You would get beat by your husband and not be able to go to school. That's the reality of it. [...] There's this whole mentality of, oh, I want to live in the past. I want to live in the past. The past was so beautiful and romantic and blah, blah, blah, blah, blah. [...] [They are] romanticizing things that just aren't real. And so they get attached to these songs that represent those times, I guess. (Kellin Hanas, interview)

The popularity of a specific jazz repertoire on TikTok is cited by the interviewees as a significant contributing factor to its success. This popularity can be observed as manifesting in certain geographical regions and social milieus. The song "Fly Me to the Moon," which, according to the corpus analysis, is one of the most popular jazz standards on TikTok, was referenced on multiple occasions in the interviews. It is "probably the most popular jazz song that a lot of people know," says Rachel Chiu, and songs from the Great American Songbook are familiar to many people who have no specialized knowledge or academic background in jazz. Rachel is referring to jazz standards like "Autumn Leaves" and "Misty," which she herself associates with singer Ella Fitzgerald and which she believes are very popular among different groups of listeners:

> I like the older songs, like, Ella Fitzgerald, like, "Misty" or "Autumn Leaves," like, typical ones. And those are usually the ones that other people have also maybe heard. If they've heard jazz, those are the songs they heard. So if you're not a jazz student, if you heard any jazz, those would be the songs. My goal is to get some people to be, like, oh, I actually know this song, although not many people know it, right? (Rachel Chiu, interview)

In addition to Ella Fitzgerald, Frank Sinatra was frequently referenced in the interviews, in part due to the fact that he recorded what is arguably the most well-known interpretation of "Fly Me to the Moon" in 1964. As asserted by Rachel Chiu, it is arguably one of the most celebrated compositions in the history of jazz. When asked whether she believes this phenomenon is exclusive to TikTok, Rachel responds with a qualified affirmative, noting that this particular song and analogous repertoires and musicians, such as Frank Sinatra and Ella Fitzgerald, are likely to be perceived as occupying a relatively elevated status, particularly within the North American cultural milieu:

> I think it's beyond TikTok, but maybe I could rephrase by saying it's one of the most popular songs in North America. Because, you know, a lot of people I know know this song. If they do know a jazz song, it would be "Fly Me to the Moon." Name a jazz song, they would say: Frank Sinatra, "Fly Me to the Moon." Or they know who Frank Sinatra is. One of the biggest jazz artists that is common and that people know. Frank Sinatra and Ella Fitzgerald, those are two big ones. But also it's big because during Christmas, Frank Sinatra sings a lot of the songs that we listen to. And so that's why I think people know this song. It's because Christmas is also one big reason why people listen to jazz, which is sadly the truth. And so I think because Frank Sinatra sings a lot of those Christmas songs, you hear his voice a lot. And so people are most likely to listen to him if they really are to listen to jazz. (Rachel Chiu, interview)

The proposition that jazz is synonymous with Christmas music is antithetical to the classification of jazz as art music, a designation that is prevalent within academic discourse. This illustrates the fact that the perceptions of content creators who are successful with jazz on TikTok – and presumably users who watch jazz videos on the platform – can differ significantly from the conventional and enduring value systems of numerous jazz fans, journalists, musicians, and academics. It would appear that the particular notions pertaining to the logic of the TikTok platform are also shaped by a multitude of music-related biographical experiences that the musicians have accumulated beyond the platform. It is noteworthy that Sam Ambers, the sole non-North American among the interviewees, reports that he did not anticipate a significant response when he posted a video of himself singing "Fly Me to the Moon" on TikTok (Ambers 2021b):

> "Fly Me to the Moon" is probably – I'm trying to think, it probably is the most popular jazz standard of all time. Potentially, potentially. I'm not sure, that's

> a big claim. But I mean it's one of the top standards, isn't it? But again, you know, whilst I like to listen to Frank Sinatra and Ella Fitzgerald, I didn't think that "Fly Me to the Moon" would resonate with the TikTok audience. Therefore, I had to try and add a bit of spice to things. Again, "Feeling Good." Whilst "Feeling Good" is probably a bit more mainstream than "Fly Me to the Moon." Again, it's not completely – it's not a pop song. So it was about adding a different element just to make it a bit more entertaining. (Sam Ambers, interview)

For the British musician Sam Ambers, the success of Michael Bublé's version of "Feeling Good" on TikTok is more evident than that of "Fly Me to the Moon." As he states, he had to make the Sinatra song "more entertaining." However, his interpretation of the Bublé song, as evidenced by his first viral TikTok video, is also shaped by the distinctive comedic stylings of the song itself. This is why, at first glance, there appears to be no notable contrast in his approach to the two aforementioned songs. Given Michael Bublé's considerable commercial success in the British music market, particularly during the 2000s and 2010s, it is plausible that Sam's assertion that Bublé's music is more aligned with pop than that of Sinatra is influenced by his personal experience and background. This illustrates that the success potential of specific musical repertoires among musicians on TikTok is also influenced by factors that are not directly related to the platform's intrinsic characteristics.

The musicians who participated in this study are acutely aware of the fact that the TikTok-specific repertoire predominantly reflects a distinctly North American perspective on jazz. Caity Gyorgy, for instance, makes this observation: "I think these songs are popular in general. 'Fly Me to the Moon,' everybody knows that song, especially – at least in Canada and the U.S., most people know that tune." Similarly, Frank Sinatra is widely recognized in North America and can be, according to Caity, described as an "iconic" figure. Nevertheless, the immense popularity of "Fly Me to the Moon" has also resulted in the song being perceived as "extremely over," as Erny Nunez argues: "I think it's overrated in North America." Caity Gyorgy maintains that the emphasis on musicians like Frank Sinatra and their most celebrated compositions ultimately reflects a certain North American centricity:

> I know I'm being very, like, North America centric right now, and I think about all of this stuff because that's all I've known. But a lot of people over North America know those songs very well from either pop culture references or

just, you know, from – it's funny, it's a part of the culture, like, Frank Sinatra is part of the culture here. (Caity Gyorgy, interview)

6.6.3 The Influence of Popular (Mass) Media on TikTok

Caity's reference to "pop culture references" indicates the potential influence of other factors on the jazz repertoire on TikTok, as identified by multiple interviewees. These factors include popular (mass) media formats such as movies and TV shows, as well as computer games. It would be misguided to assume that digital platforms are entirely free from the influence of classical mass media. Indeed, there is evidence to suggest that the content and formats of such media are sometimes transferred to the platform context and transformed according to certain platform conventions (Jost 2022). In the case of *The Fairly OddParents*, Brooklyn Stafford has argued that the presentation of theme songs from such formats on TikTok constitutes a strategy with a proven track record of success. This is due to the fact that these songs, as Brooklyn argues, evoke feelings of nostalgia in a significant proportion of users (see above). The question of whether TikTok users engage in nostalgia when listening to such songs, or whether their individual associations are more diverse, is irrelevant in light of the undeniable fact that content disseminated by mass media plays a significant role in the popularization of certain songs on the platform. The song "It's Been a Long, Long Time" is a compelling illustration of this phenomenon. The corpus analysis indicates that it is one of the most prevalent jazz compositions on the platform, with nearly all of the jazz musicians in the top one hundred having uploaded one or more renditions. The song achieved a notable degree of popularity among younger demographics following its inclusion in the closing sequence of the 2019 film *Avengers: Endgame*. This cinematic work is based on the superhero comics *The Avengers*, published by the U.S. company Marvel Comics. It is evident from the responses of the interviewees that the popularity of this musical piece on TikTok can be attributed to the popularity of the films, particularly given that they are sometimes able to recall the corresponding film scene. In this context, Caity Gyorgy highlights the popularity of Marvel movies in North America:

> [Regarding] "It's Been a Long, Long Time" [...] I have speculations. This one is popular but I think it really got a resurgence because it was at the end of one of the *Avengers* movies, recently. [...] Where [Captain America] goes back in time and he lives out his life with agent Peggy Carter and they grow old

> together. [...] The *Avengers* movies are huge. The Marvel movies are crazy big over here. [...] The last scene is him dancing. It's, like, the 1940s or something. At least it looks like that from the outfits that, I think, are just from the time period based on when Captain America is from. This last scene is him and agent Peggy Carter, who's his partner, dancing. And that's the song that plays into the credits. And so that one is, like, a huge earworm I think for people as well, especially if people are a fan of the Marvel *Avengers* movies. So those are very popular songs. (Caity Gyorgy, interview)

This appears to corroborate Kellin Hanas' characterization and critique of the way in which younger individuals display a penchant for referencing and romanticizing the era spanning the 1920s to the 1950s. Not only does the prevalent jazz repertoire on TikTok originate from this period, but on occasion, the mass media formats that inform the repertoire on the platform unmistakably evoke this historical epoch. According to Erny Nunez, this music now has a "very, very young fan base." He was astonished to discover that the song "It's Been a Long, Long Time" had unexpectedly gained immense popularity among the younger demographic as a result of the *Avengers* film: "After that I started seeing a lot of young kids around my school sing it when the movie came out and I was, like, I'm seeing other teenagers at my age singing this type of song. But I do think having big types of people, like, listen to this music and play is very impactful on this young generation."

As Erny Nunez observes, video games can also serve as a source of inspiration and influence for the jazz repertoire on TikTok. In one of his most popular videos, Erny performs the 1941 version of the song "I Don't Want to Set the World on Fire" by the vocal group The Ink Spots (Nunez 2021). The song is featured on the soundtrack of multiple versions of the video game series *Fallout*, which served as a source of inspiration for Erny. "I played the video games, actually. I played the video games ten years ago. I was, like, this game has a lot of really old music that I should probably try. And so I think that was the first song that came into my mind that I should sing on TikTok." Kellin Hanas also emphasizes the impact of such media on the jazz repertoire on the platform and presents an analysis of another video game with a comparable soundtrack: "That's a really big thing. There's this video game called *BioShock*, and the whole soundtrack of that game is really, really old tunes. A lot of big band stuff and a lot of old vocal stuff."

Furthermore, it appears that a considerable number of young individuals are first exposed to specific jazz compositions via their use in Netflix series soundtracks, as Stella Cole has postulated. After uploading her rendition of "Fly Me to the Moon" (no longer available on TikTok), she was taken aback by the multitude of comments that referenced the South Korean Netflix series *Squid Game*, which features the aforementioned song on its soundtrack:

> When I posted that video of "Fly Me to the Moon" and it got, you know, maybe 300,000 views or something, so many people were commenting, like, oh, this is that song from *Squid Game*. And I was, like, what? I had never seen *Squid Game*. And I was, like, what are people talking about? This is not from a show that was made two years ago. This is one of the most famous songs in America. (Stella Cole, interview)

6.6.4 Jazz and Humor on TikTok

TikTok is occasionally dismissed as a platform for juvenile content that may be challenging for older users to understand. While such critiques often have an undertone that smacks of a blanket criticism of contemporary culture, it is undeniable that humor is a pivotal aspect of the platform. The short-form video format seems to lend itself well to presenting short skits, and musical content in particular is often used to support punchlines or the narrative of a video as a whole (Abidin and Kaye 2021, 58). It is therefore unsurprising that jazz is sometimes associated with humorous content on the platform. As the corpus analysis demonstrates, such content is a common, if not dominant, facet of jazz on TikTok. Furthermore, several of the interviewees have also produced videos with humorous connotations or even a focus on humor, most notably Kellin Hanas, Caity Gyorgy, and Sam Ambers.

The integration of humor and jazz was met with a favorable response in the interviews, with particular commendations directed towards Kellin Hanas's videos. As Caity Gyorgy argues, Kellin succeeds in combining jazz with humor in a way that does not denigrate the music. Rather, it offers an opportunity to introduce the technically proficient aspects of jazz to those who might otherwise be unable to access this genre: "Kellin does an incredible job of blending comedy with music. [...] I think that there has to be a way to make the virtuosity accessible. And I think, like, with Kellin combining comedy with the virtuosity is a really great way for people to be, like, oh, that's hilarious." Kellin Hanas's own sentiments align with this viewpoint. At the inception of her TikTok career, she did not upload any content related to music, instead focusing on comedy. However, over time, she conceived of a novel approach that integrated both genres. According to Kellin, it is important to note, however, that not all content related to jazz would be recognized on the platform, as most of the audience does not have a strong connection to this genre of music. That is to say, the content would have to be tailored accordingly, according to Kellin:

> The general public doesn't really understand this music. And so I was, like, okay, what if I took my trumpet and did something just stupid? Just something so random that wouldn't make any sense? [...] Some people will think that I have a book I write my ideas down in and – no. Literally every time I make a TikTok, I go: I want to make a TikTok, what would be stupid? And then I get the idea within 0.5 seconds and I'm, like, okay, I'm going to do this because this seems stupid and dumb. (Kellin Hanas, interview)

While Kellin discusses the perceived frivolity of the punchlines in her videos, most TikTok users appear to be drawn to the distinctive blend of comedic content and Kellin's trumpet performances. This is also evident in the commentary accompanying Kellin's videos. "People were commenting on the playing part. And I was, like, oh, I can do something with the playing now and also not compromise me wanting to be, like, goofy. I can do both at the same time." In her discourse, Kellin describes how she discovered a strategy or a distinctive selling point that can be repeatedly implemented. She stresses the necessity of maintaining the prominence of the musical component despite the incorporation of humor. The integration of jazz and humor on TikTok represents a challenging balancing act for musicians. Kellin asserts that she is careful not to convey the impression that she is disparaging jazz on her TikTok profile. At the same time, the nature of TikTok is inherently conducive to the presentation of humorous content, particularly for members of Generation Z (those born between 1997 and 2012), who have become accustomed to such forms of expression. Kellin elaborates:

> It's a difficult thing to balance because [...] I mean, this music is serious. [...] None of my videos are saying: Oh, this music is dumb and silly. [...] I'm not compromising the music when I do it. I'm making sure that the music is at a high level to be enjoyed at. To give respect to the music part. Obviously, if I put on a jazz song and made it stupid, that would be disrespectful. [...] But, you know, that's something that I've been worried about. [...] I care about it, obviously, but [...] there's a ton of creators on that app that do everything based off of humor. [...] I'm part of Gen Z. I think that we are just the silliest. Our humor is so weird and we are – our generation really relies on humor, and

> so combining, you know, something that I'm good at and that I enjoy doing is kind of a little unique. You know, there's a lot of jazz musicians, but most people in the world aren't a jazz musician, let alone a female jazz musician. Combining kind of the shock factor of that plus something that relates to the younger audience that's on this app – because they're just looking to laugh, like, we're all looking to laugh at something just ridiculous together because that's what our generation thrives off of. So, it's just – it's like a hard balance. [...] I think I just got really lucky that people liked it because it could have easily gone the opposite way of, like, this is annoying. But I think that I did it in a way so that it's paying respect to the music and combining something that the younger generation likes. (Kellin Hanas, interview)

Sam Ambers asserts that jazz videos with a humorous connotation can also be appreciated by jazz musicians. He cites one of his most popular videos, a nine-second clip captioned "an everyday occurrence" (Ambers 2022a), in which he is seen with another singer in front of a doorway, as an example. The video incorporates a text layer that provides a description of the depicted action: "POV: You walk past 2 jazz musicians having an argument." The action is confined to the two vocalists engaged in a verbal altercation, employing scat singing as a means of insulting or competing with one another. The camerawork employed is intended to create the impression that the situation has been captured spontaneously and unobtrusively, within the context of everyday life. Sam Ambers states that the video was created spontaneously, in just a few minutes to spare during a meeting with two other TikTok content creators. The video rapidly gained considerable online attention, even beyond the TikTok platform, according to Sam:

> The thing is, that video got reposted by a lot of jazz memes pages on Instagram. And the funniest thing is, when I was at the Jazz Conservatoire last week, people didn't know me from the Samsung videos. They knew me from that video, which is a funny thing. They were, like, oh my God, he's the guy from the jazz video. And that was quite funny. (Sam Ambers, interview)

As exemplified by the jazz meme websites referenced by Sam, the conjunction of jazz and humor in digital spaces is not an anomalous phenomenon and it does at times attain a considerable degree of popularity. The observation that the majority of Sam's videos, as he himself attests, are not widely known among jazz students indicates that his content reaches an audience of listeners with an affinity for this music who are not professionals in the field of jazz. It can be reasonably inferred that TikTok is reaching a distinct subset of the jazz audience that is not necessarily engaged with more traditional media channels.

6.6.5 Virtuosity

In the case of Kellin Hanas's videos, as previously stated, multiple interviewees highlighted the distinctive combination of comedic elements with her relatively advanced technical proficiency on the trumpet. Her trumpet playing has on occasion been described as virtuoso, which is undoubtedly a contributing factor to her success. However, the results of the corpus analysis indicate that musical virtuosity is not a significant factor in the field of popularity peaks of jazz on TikTok. Nevertheless, virtuosity is indubitably a pivotal element of the aesthetic value system of numerous jazz musicians and aficionados, particularly with respect to the improvised solos that are of paramount importance in jazz. The interviews addressed the concept of virtuosity and the notable absence of virtuosic elements in the videos posted by the most popular jazz musicians on TikTok. There is a general consensus that it is difficult to convey virtuosity in a short-form video. This is exemplified by Kellin Hanas, who has demonstrated the necessity of presenting virtuosity in a manner that is accessible to a broad audience. Caity Gyorgy has postulated that the logic of the platform itself encourages musicians to refrain from displaying virtuosic abilities:

> I think [...] virtuosity is typically demonstrated in a longer setting. [...] I think it's very easy for somebody to sing an incredibly high note, you know, once and record it and post it. But to be able to do it non-stop for an extended duration of time and just prove that you have worked on this and prove that you've developed your instrument, you've developed your skills and your understanding of theory and harmony enough to, you know, automatically, you know, do a whole thing. [...] There are some people that are just fantastic musicians and, like, you know, like, they're virtuosic, and maybe they just don't show their virtuosity. (Caity Gyorgy, interview)

Furthermore, musical virtuosity or complexity can act as a deterrent for a significant proportion of users. It is therefore essential to provide content that is relatable for the largest possible number of users. As Sam Ambers argues, this implies that even videos that are perceived as not being produced to a high standard can be extremely successful: "Ironically, the lower production or the lower quality videos tend to do better because I think on TikTok people want relatability, you know. I don't think they necessarily want virtuosic playing that, you know, doesn't seem attainable or realistic."

Conversely, Caity Gyorgy's videos, which prioritize showcasing her technical abilities, have also garnered admiration from numerous TikTok users. This is particularly evident in several videos in which she scats at high speed, primarily singing instrumental jazz solos that she has transcribed beforehand. In reference to the previously discussed video captioned "Can you tell which song I'm soloing on? Bonus points if you can count how many times I breathed!" (Gyorgy 2021a), Caity recalls users' reactions and her own amazement at the video's popularity:

> I was, like, well, I don't have a TikTok for today. So I guess I'll post this one. And it blew up. [...] There were so many people being, like, I have no idea what you're doing, but I know it's impressive. And so I reach this audience that was, like, not jazz audience at all, but they were very impressed by it. Which is kind of interesting to me and – yeah, I'm not really sure what to think about this one because I was just sort of, like, blown away that people actually liked it, because it was just – I thought it was extremely niche content, but apparently not. (Caity Gyorgy, interview)

6.6.6 Original Compositions

The corpus analysis revealed that original compositions do not occupy a prominent position within the context of jazz on TikTok. The interpretation of jazz standards is a common practice among jazz musicians, which may explain why original compositions play a less crucial role in this field than in many other popular music cultures. Nevertheless, creativity, particularly in the context of improvisation, represents a pivotal aesthetic criterion in the domain of jazz. Additionally, the act of composing original pieces represents a pervasive area of engagement for numerous jazz musicians. Of the musicians interviewed, only two – Stacey Ryan and Caity Gyorgy – regularly perform their own compositions in their videos on TikTok. Caity Gyorgy acknowledges that she did not initially anticipate that her own compositions would achieve success on the platform. However, she experienced a profound sense of elation upon observing this outcome: "I started posting my original songs and then those were the ones that have done the best. And that was just sort of, like, very incredibly encouraging for me because I thought, you know, like, people want to hear my original music."

The musicians' reluctance to upload original compositions on TikTok is occasionally motivated by strategic considerations. Brooklyn Stafford, for instance, notes that she has been composing her own music for an extended period but has not yet shared it on TikTok due to her desire to expand her audience and to reach "enough people." Kellin Hanas does not believe that TikTok is an optimal venue for showcasing her original compositions:

> It's just not the place, you know? I do compose. I have compositions that I've played and I just don't think TikTok is the place to put that because it just wouldn't reach an audience because the audience is just not there. On my Instagram, maybe, you know, on my professional Instagram specifically for my music that people are following to see that stuff. But [on] TikTok, half of the people are there [...] just because of the other stuff that I do, other than trumpet. So I don't think anybody would be interested in seeing a three minute, four minute long video of an original that I wrote because that's just not what they're there for, you know, they're there to see some crazy, weird trumpet thing and then something else unexpected happens, you know? (Kellin Hanas, interview)

6.6.7 Jazz on TikTok vs. Jazz Beyond the Platform

The sections of the interviews that address questions of virtuosity and original compositions indicate that the affordances of TikTok's platform logic not only encourage musicians to prioritize a specific repertoire, as evidenced by the popularity of Great American Songbook songs on the platform, but that – based on their experience on the platform – musicians also refrain from presenting certain facets of their musical work on TikTok. In this manner, the affordances of the platform and the musicians' algorithmic imaginaries not only impact the selection of which musical repertoires are performed, but also promote the exclusion of certain styles, thereby preventing them from being visible on the platform.

Some interviewees note that the jazz music observed on TikTok may differ from traditional representations of the genre in other contexts. They attribute this discrepancy to the distinctive logic of the TikTok platform, which necessitates that musicians capture users' attention immediately. Brooklyn Stafford argues: "Jazz music in general on TikTok and social media is so much different than what's popular, I would say, in the real-world setting, and what catches people's attentions." Stella Cole suggests that the conjunction of TikTok and jazz is a tenuous one, particularly in a live setting. She argues that jazz is predicated on the immediacy of interaction between musicians, encompassing a mode of situational and spontaneous musical creation within a collective and that the situation on TikTok is, in fact, the opposite, which constitutes a significant challenge for her as a musician:

> It sometimes is difficult because music and especially jazz music is supposed to be this very, like, in the moment thing. [...] Everybody's playing exactly what they're feeling and, like, improvising accordingly. And it's supposed to be a very, like, of the moment inspiration. And it's interesting when you try to put an art form like that onto a social media app [...] that assumes that people have very short attention spans and that prioritizes things that are going to be, like, a quick and sort of instant source of pleasure. So that's definitely an interesting thing to navigate. (Stella Cole, interview)

She sometimes considers the music she creates on TikTok to be a distinct form of musical expression that is largely unrelated to her live performances:

> A jazz tune, you know, it's going to be at least, like, four minutes long once the band has all taken a solo. That's such an important part of it. And having that when I'm performing in real life and then trying to go back to TikTok and film something where I'm just sort of, like, smiling and singing 30 seconds of a song – it's definitely different, it feels like a different art form. (Stella Cole, interview)

Stella states that the platform logic of TikTok exerts a pivotal influence on which facets of jazz gain popularity on the platform and on her own self-perception as a musician: "In the end of the day, a computer is what's deciding who's successful on TikTok and who isn't. And that changes the art, and it changes how it feels to be an artist."

It should be noted, however, that the computer itself does not determine success or failure on TikTok. Rather, socially prevalent stereotypes play an important role in the socio-technical negotiation of visibility on TikTok, especially in relation to the construction of beauty and regarding certain body norms.

6.7 Gender Stereotypes, Body Standards and Beauty Norms

The findings of the corpus analysis indicate that the platform logics of TikTok and the socio-technical interactions between the platform and its users and content creators exert a significant influence on the visibility of jazz repertoires on the platform. Moreover, they also shape the potential for success of certain (groups of) musicians on TikTok. As has already been demonstrated, this phenomenon is particularly prevalent among relatively young, female, *white* individuals between the ages of eighteen and twenty-five who conform to specific conventional beauty standards. These ideals pertain, on the one hand, to the physical appearance of the musicians and, on the other hand, to how they present themselves on TikTok. Upon first observation, it would seem that female jazz musicians in particular attach significant value to their appearance, encompassing elements such as makeup, hairstyles, and attire. While such processes are not exclusive to TikTok but are a ubiquitous feature throughout the entertainment industry (and beyond), the interviews demonstrate that a distinct *beauty action* (Degele 2006) is particularly evident on TikTok. The algorithmic imaginaries of the musicians are of significant consequence in this regard, at times exerting a decisive influence on their actions with regard to how they portray themselves on TikTok.

In the course of the interviews, all the musicians emphasized that, from their perspective, certain normative ideals of beauty play an important role on TikTok. In particular, they noted that conforming to a certain physical ideal is sometimes described as one of the basic requirements for success on the platform. These assumptions are typically derived from prolonged observations of the platform's logic. In certain instances, musicians have specific ideas about the best way to present themselves on TikTok. These ideas then become guiding principles, influencing the image of jazz that emerges on the platform to a similar degree as ideas about promising musical repertoires and musical features. Caity Gyorgy provides a summary of the relevance of certain beauty ideals based on her own experience:

> If I don't wear makeup, my videos don't do well. If my hair is, like, up in a bun, it doesn't do as well. [...] In a lot of my videos, you know, I've got, like, the full makeup on. Like, sometimes, if I'm making a TikTok, I will go and I put my fake eyelashes on. I glue them on, and just so that I can do the TikTok and that it will do well. [...] They do better when you look better. [...] The ones that aren't doing as well are the ones where I'm not really wearing makeup.

> Or, yeah, I had my hair in braids for one. Didn't do as well. [...] It's definitely a factor. If I do my makeup, they do better. If I don't, they don't do as well. (Caity Gyorgy, interview)

Identifying promising patterns is of great importance, both in terms of which repertoires are performed and which specific techniques (or tricks) are employed to make the TikTok videos more engaging. This process also has a significant impact on the way content creators present their own bodies. This is evident in numerous other interview passages that are strikingly similar in content to the above statement by Caity Gyorgy. For example, Stella Cole emphasizes: "It's definitely something that I've noticed. Videos perform better if I look better. I've posted some of my, like, musically some of my best videos, I think. And they haven't done well because I maybe wasn't close up enough to the camera so people could see my face or I wasn't wearing makeup." It is evident that a particular visual presentation appears to increase the probability of success. As Brooklyn Stafford observes, creators must acknowledge this logic of success to a certain extent, which she is willing to accept:

> Sometimes you do have to play the game. And if you have attributes that work, you have to milk it, right? If you know that pretty sells, if conventionally attractive sells, then: hey, if it's working for you, then do it, right? Like, you have nothing to lose. [...] Just the reality of what TikTok wants to push is conventionally attractive. That's just what I've seen. It's stupid. But, you know, that's just the weird world we live in today, you know? (Brooklyn Stafford, interview)

Such assumptions are not arbitrary; rather, they are based on concrete experiences made by musicians with how they appear on the platform and the corresponding reactions of TikTok users. To illustrate, Rachel Chiu describes an incident during a live stream on TikTok in which she altered her attire and subsequently observed a notable surge in the number of views:

> One time I was singing "All My Life," and I don't know if you've ever seen the videos that are, like, get ready with me, where [...] these girls are doing makeup, and then they're singing. So I've been doing that where I was just singing, I was getting ready for a banquet at my school. I started off the video without any makeup on. I was wearing, like, this old t-shirt that I had. And by the end of my live stream, I went to the side and I changed into my dress, and I asked for advice: Oh, which earring should I wear – while I was singing.

> And I noticed that the number of views went up by so much more after I had my makeup on, my hair done, my dress on. I had way more views than I did when I had that old t-shirt on and no makeup. So I think it does affect the level of success in the video to a certain degree. (Rachel Chiu, interview)

Although Rachel is describing her experience in the context of a live stream and not in the short-form video context typical of TikTok, the huge relevance of beauty ideals and body standards on the platform can nonetheless be observed. Moreover, it is possible to draw conclusions from this interview passage about how algorithmic moderation contributes to the perpetuation of normative beauty ideals. In order to reach a larger number of users with her live stream, Rachel's video had to be featured on the For You pages of as many TikTok users as possible. The abrupt surge in views subsequent to her outfit change indicates that her live stream was algorithmically prioritized, subsequently garnering the attention of numerous users. In other words, the TikTok algorithm was required to take the initiative. With regard to the design of short-form videos, certain learning processes also occur over time with respect to how content creators stage their own bodies. For instance, Stella Cole illustrates how, over an extended period, she discerned which staging techniques were most conducive to success, subsequently designing her videos in accordance with these criteria:

> I would definitely look at what videos did well and I'd be, like, okay, I'll keep doing that. So I figured out that people or the algorithm, whichever it is, really like when I – it's, like, a close-up of my face. [...] I'm a very smiley person, so I'm usually smiling anyway. But people are always, like, oh, your smile is so nice. It makes me feel so happy. And so usually: pretty singing, sort of happier songs, smiling in my videos. (Stella Cole, interview)

At this juncture, it becomes evident that the musicians' conceptualizations of potential pathways to success are largely informed by both the platform algorithms and the user reactions on TikTok. This underscores the fact that their imaginaries extend beyond the algorithms themselves, as the term algorithmic imaginaries would suggest. Instead, they encompass the broader events and interactions that occur on the platform. As Stella's statement illustrates, even successful content creators are unable to ascertain the extent to which algorithms, users, or their interactions are responsible for specific processes on the platform.

The experiences of the musicians interviewed indicate that beauty ideals and body standards play a key role not only on TikTok, but also on other digital platforms. Caity Gyorgy, for example, shares her experiences with Instagram: "I know that the videos where I have my makeup done do a lot better than the ones where I don't. [...] I noticed that years ago on Instagram. You know, when you post a video, when you've got a full face of makeup, people like that one more than the one where you don't have makeup on." This phenomenon, observed on social media platforms over an extended period of time, appears to manifest similarly on TikTok. Furthermore, it can be argued that the prevalent attitudes and behaviors in society at large are reflected and negotiated within digital spaces. For example, the prioritization of so-called *standard beautiful bodies* is not exclusive to digital platforms; however, it may be intensified in digital spaces. Ultimately, according to Brooklyn Stafford, the importance of appearance on TikTok is on par with that of musical performance:

> The visual aspect is so important. It's almost as important as the content you're playing. And people find their niches too, right? You don't have to be, like, freaking – I don't know – Bella Hadid to be successful on TikTok. But being presentable, like, looking your best is very important, too, right? [...] At the end of the day, [...] when I look good, it definitely helps. It never fails to help to look good in a video. (Brooklyn Stafford, interview)

Nevertheless, the musicians frequently assert that decisions about their performance on TikTok are ultimately based on their personal preferences and are therefore a matter of self-determination. Caity Gyorgy, for instance, maintains that the authority to make such decisions resides with her alone. She further states that she does not wear makeup with the intention of aligning her appearance with the expectations of strangers on the platform: "When I'm making my TikToks, I know that the videos do better when I have my makeup on, and that's fine. But I'm not putting my makeup to help other people, you know, like, not make comments on my videos. [...] I've had social media since I was 14, I've always put on makeup, you know." In essence, Caity's argument is that the decision to leverage physical attributes for the purpose of enhancing one's success in the TikTok content creator realm is ultimately at the discretion of the creators themselves.

Concurrently, several interviewees asserted that TikTok users have specific expectations regarding their physical appearance and that they occasionally feel compelled to align with particular beauty standards. It bears pointing out that this phenomenon is not exclusive to TikTok. Caity Gyorgy, for instance, offers insights into the responses of her Instagram audience to videos in which she is not wearing makeup: "[That] happened on my transcription account on Instagram. I posted transcriptions where I don't wear makeup, and I've had people saying, like, you look sick, are you okay? And I'm, like, yeah, I'm fine. I just don't have an even skin tone." In the context of TikTok, Rachel Chiu notes that users have certain expectations regarding the appearance of content creators:

> I think TikTok really exposes that, and I think that's where it's a little difficult because people on there – you can't control what they say, but they can be quite judgmental. And as a woman you feel like you have to be presentable. And if you're singing, like, jazz music, you need to look very, like, gentle and fair, you know what I mean? So I think there are definitely unspoken expectations for women. (Rachel Chiu, interview)

Similarly, Caity Gyorgy articulates a perspective that is, at its core, at odds with the aforementioned expectations. She states that these expectations are fundamentally incompatible with her personal values:

> I think there's an expectation for people to look good, and for people's videos to do well, it's sort of expected that they put effort into their appearance, as well as the content. I think the appearance sort of factors into the quality of content in some ways. Not that I believe in that. It shouldn't be a thing, of course. But I think that the appearance does definitely affect how people perceive the quality of content. (Caity Gyorgy, interview)

It is evident that the musicians are under considerable pressure to justify their actions. While young women are particularly influenced by the platform logic of TikTok to conform to the conventional standards of beauty, it is crucial for them to highlight that such performance practices may not align with their personal ideals.

Nevertheless, it is not possible to circumvent this subject entirely. As previously stated, musicians frequently peruse the comments appended to their videos in order to better understand users' preferences and expectations. Additionally, comments are frequently directed at the musicians' appearance. Stacey Ryan's video for "Boogie Woogie Bugle Boy" (Ryan 2020), previously referenced, serves as an illustrative example:

> You can't help but notice if you're wearing maybe a lower cut shirt or makeup, you do get the comments. [...] On the "Boogie Woogie Bugle Boy" video – I remember because it was all older people, right, who are seeing it or a lot more because it was shown to an older audience. And I remember getting one or two comments being, like, wow, I really like your shirt, with, like, a winky face or something. And I, you know, I was, like – I don't think these comments are inevitable, but you definitely notice them, you know, when you get them, they definitely stand out for sure. (Stacey Ryan, interview)

In such circumstances, the musicians interviewed may also feel compelled to justify or even apologize for a certain look or appearance in their videos if it does not conform to the prevailing beauty standards. As a case in point, a video by Stella Cole was viewed and discussed with all the musicians as part of the interviews. This was done in order to facilitate a discussion about the role of beauty standards on TikTok with the interviewees. In the video captioned "Reply to @_amyjhs where my Julie London stans at?" (Cole 2020), in response to a user request, Stella provides a reply within the comment field that is visible within the video frame: "Please can you sing some Julie London 💗." Stella then performs an excerpt from the 1953 song "Cry Me a River," written by Arthur Hamilton and made popular in 1955 by Julie London. In regard to the discourse surrounding beauty standards on the platform, Stella's concise verbal statement in this video is of paramount importance. "Yes, I am in pajamas. Yes, my hair is in two braids. Yes, I'm filming a TikTok." This video does indeed deviate from the majority of Stella's other TikTok videos in terms of its staging. She is wearing a relatively loose hooded sweatshirt, with no discernible makeup, and her hair is pulled back with a headband and braided at the sides. Furthermore, the video is not illuminated with a ring light or other professional lighting, which results in a relatively dark overall appearance. There are striking contrasts to most other videos, where Stella can be seen wearing makeup, her hair loose and styled, and well lit.

After viewing the video in question, all of the musicians interviewed proceeded to debate the extent to which beauty standards play a pivotal role on TikTok. They noted that, for those who have achieved a certain level of success in content creation, there is an inherent pressure to apologize to users if their appearance does not align with the prevailing body standards and beauty norms. In response, all of the interviewees corroborated the relevance of beauty standards, and in some cases, they recounted their personal experiences with the previously described pressure to validate themselves: For example, after viewing Stella Cole's video, Rachel Chiu went on to show the interviewer one of her own videos, which had not previously been uploaded to TikTok and bears a striking resemblance to Stella's video. Rachel then proceeded to describe the background of the video in question:

> I haven't posted this, that's why you can't see it [...] I was basically saying in the video: [...] I have acne everywhere. I'm not wearing any makeup. It's 3 am and I'm so tired. I just finished school. I look like a mess. I guess that's a little bit similar to the video you showed me. So I think we are as women, like, pretty aware that we have to look presentable. (Rachel Chiu, interview)

The sentiment articulated by Rachel on the necessity of aligning with specific beauty standards on TikTok, particularly as a woman, resonates with numerous musicians interviewed for this study. Stacey Ryan also describes how she occasionally feels compelled to justify her physical appearance to TikTok users:

> I remember posting a couple of times and being, like, oh, so I look terrible. [...] I had a sunburn once and I was, like, I already know I'm sunburnt, you don't have to tell me. You do feel the need to justify why you look the way you look if it's not good, which is terrible. [...] You just feel the need to make sure they know that [...] you don't look like that every day or you just really wanted to post a video. Like, a lot of people, you see they're, like, oh, I look like crap, but I just love this song and I feel, like, that's a reaction that we all have definitely to posting a video where you maybe don't look your best or don't feel your best. (Stacey Ryan, interview)

However, the musicians are not only responding to the perceived expectations of the TikTok audience, they are also responding to the expectations that are sometimes placed on them by people in the music industry. Stella Cole explains that her video – the one that was discussed with all the musicians interviewed for this study – is part of her strategy for dealing with these expectations and to sometimes deliberately subvert them:

> You've seen probably a couple of times me being, like, sorry, I'm not wearing makeup, but I'm filming a video anyway. Because it just feels – I don't know why I would apologize for it. That's kind of messed up, honestly, but it just feels to me – like I was saying earlier. In my experience with putting myself singing or acting on video, it's important to look, like, put together. And I guess to me, put together means that I have my hair and makeup done. And

> then people, certain people in the industry, like managers or producers or whatever, who have talked to me, are, like, it's important that on your videos you look—you have this sort of, like, Marilyn Monroe, Elizabeth Taylor energy, which is code for saying, like, you have to be extremely beautiful at all times and make that a part of your brand. You have to have sort of well-done hair and the red lipstick, and that's what people want to see. They don't want to see you in a t-shirt. So I guess me apologizing for not wearing makeup is me acknowledging, like, yeah, I know you guys want to see me, I know how you want to perceive me, and I'm not going to do that right now, but I'm still in the mood to sing. (Stella Cole, interview)

The interviewees do not always wish to meet the expectations placed on them, although they frequently defer to the logic of success in the context of TikTok, which is influenced by beauty ideals and body norms. Videos like the one by Stella Cole discussed here represent a minority of cases. Some musicians question the value of posting videos without staging their bodies in a particular manner. Rachel Chiu offers a perspective on this issue: "Sometimes I want to record a video but I'm, like, [...] I'm not wearing makeup or I'm not wearing something that I should be wearing for a video, right? I'm wearing my pajamas or whatever." Furthermore, Stacey Ryan notes that her TikTok-related production practices are occasionally shaped by the expectations associated with her staging. She asserts that she occasionally refrains from recording videos if she does not feel adequately presentable:

> A lot of the time a TikTok inspiration comes, like, it's nighttime and you're already in your pajamas. I feel, like, you shouldn't let that hold you back. And a lot of people it doesn't. But then again, you start thinking in your head: Oh no, what if I have this amazing idea and I sound really good right now, but I don't look good. And then you don't make the TikTok. (Stacey Ryan, interview)

It should be noted that some interviewees raised these issues proactively, prior to being specifically questioned about them during the course of the interviews. This was particularly prevalent in the interviews with female musicians, although the intensity of this varied.

However, it is similarly crucial for the interviewees to assert that a particular appearance is not a guarantee of success on TikTok. Once more, their personal experiences and observations of the platform's underlying logic serve as the foundation for this interpretation. In recalling her initial experiences with virality on TikTok, Brooklyn Stafford attributes a portion of her success to her appearance in the videos. However, she also emphasizes the importance of combining specific physical attributes and a distinctive look with creative ideas to achieve success on the platform:

> I kind of found that [a certain look] was important when I did – when the first video went off. [...] I had a nice crop top, little red pants, like, my abs were showing, and that went off. [...] [M]aybe it was a combination of – it was obviously a combination of looking good and having a good idea, right? You're not going to just get views because you're pretty, just get views because it's a good piece. But that's just from my experience. That first video that went off, as opposed to my first videos, which I look kind of frumpy, you know. (Brooklyn Stafford, interview)

Moreover, the interviewees stressed that, when viewed through the lens of TikTok's platform logic, it is evident that the visual presentation of content creators plays a pivotal role in determining success. This is related to the attention principles of TikTok, which stipulate that creators must capture users' attention within the first few seconds. Rachel Chiu argues: "You have two seconds to attract someone's attention before they scroll away. So, whether that be with your music, or your voice, or with your makeup or appearances. You have to find a way to make sure you grab that person's attention." Additionally, Sam Ambers emphasizes the significance of a distinctive visual presentation as a pivotal strategy for garnering the attention of both users and platform algorithms, a concept he has repeatedly highlighted throughout the interview:

> I feel, like, if I have to make sure that I'm looking as good as possible for every video, I think that would drive me insane. I don't think I can do that to myself. [...] But it's very true, though. It is very true that looks – I mean, it's about first impressions as well. You want to look nice and presentable and, you know, it does help if you're attractive. Being attractive helps tick the algorithmic boxes, basically. (Sam Ambers, interview)

It is noteworthy that one of the two male interviewees highlighted the potential challenges associated with maintaining a consistent and meticulous approach to personal grooming over an extended period of time. It is evident that the specific expectations regarding appearance on TikTok have a significant impact on the work of female musicians, although male interviewees do not deny the influence of certain beauty standards on their videos, or at the very least, on their strategic considerations. For instance, Erny Nunez typically dons a suit in his videos, a choice he regards as deliberate. This sartorial choice distinguishes him from other male singers on TikTok. Moreover, he is inherently in competition with female singers on the platform, who attach considerable significance to their appearance:

> There is a certain set of beauty standards that are proposed. And so, for me, as a Hispanic singer and probably one of the biggest male jazz singers on this app, I find that if it were just me wearing a normal t-shirt and just singing songs, I wouldn't get as much traction compared to my female counterparts because they put the time and effort to look as pretty as possible. But I think that putting the suit on is unexpected. You don't see many male people on this app just having suits on all the time. That's very rare. I do agree that beauty does have a big impact. (Erny Nunez, interview)

From Sam Ambers' perspective, physical appearance is a significant factor for male musicians as well:

> It is so important and it's genuinely something that I've been slacking on. I'm not even joking. [...] [T]he main motivation for wanting to get in a better shape or to look better is because I know that it will result in – well, I expect that it will improve results on TikTok. [...] If you see someone that's really attractive on the For You page, you're going to give them more of a chance than someone that you don't find attractive. I think that's pretty basic. [...] I've got a mate called Adam and he does music and his demographic is, like, 82% or 80% female. So he says to me, the main thing about his videos is looking good. Well, not the main thing, but a very important thing is looking good. [...] If you look good, you've got a decent chance of keeping viewer attention of people commenting, saying, oh my God, you look brilliant and people following you because you're good looking. (Sam Ambers, interview)

Although the orientation towards certain beauty standards is part of their daily business, the musicians occasionally experience discomfort with their actions on the platform or with the expectations placed on them. Some of them describe a state of frustration regarding the apparent relevance of certain beauty standards. This is largely attributable to the fact that the interviewees are acutely aware that their behavior could be perceived as opportunistic, given that their appearance evidently contributes to their success on the platform.

Stella Cole presents her perspective on the role of beauty standards on TikTok and their broader relevance:

> I'll be honest, I think it's very important and I wish it weren't. And that has been one of the most frustrating things about the platform for me, even as I've found success with it. And, you know, it's weird to be, like, yeah, I'm conventionally attractive, but I think that it's [...] easier for people who are conventionally attractive, 100%. (Stella Cole, interview)

Additionally, Stella asserts that the remarks made by TikTok users regarding her physical appearance can, at times, be perceived as disparaging. She characterizes her experiences on the platform in this context as being particularly distressing:

> Any comments on your appearance are just, like, always going to be weird. Even if it's a compliment, it's always just still going to feel sort of weird that there are hundreds of thousands of people on the internet commenting on my appearance. I posted one video from 54 Below a couple of weeks ago. It was, like, from the side, and my stomach didn't look completely flat because my stomach isn't completely flat. And so many people were, like, are you pregnant? Why do you keep touching your stomach? Are you pregnant? And I'm someone who's definitely struggled with body image and stuff like most women have. And that was hard. I was, like, this sucks, and it's definitely not fun. (Stella Cole, interview)

Stella notes that it is particularly troubling that such comments are typically made by male viewers who are significantly older than the musicians who were interviewed for this study:

> I get a lot of comments from older men and they're not commenting on the music or my voice. It's, like, oh, you're so beautiful. I want to marry you. Oh my goodness. Imagine waking up to this voice every morning, so gorgeous. That smile, those eyes, and it's, like, thank you. That's very nice. When it's not coming from old men, it feels, you know, it's, like, oh, that's a compliment. (Stella Cole, interview)

The ability to respond to negative comments is a significant challenge for many of the individuals interviewed, particularly when the comments target their physical appearance and are based on conventional beauty standards. Additionally, it can be particularly distressing when commentators make derogatory or offensive remarks. The issue was broached by nearly all interviewees, even in the absence of a direct inquiry, indicating that those who excel in content creation on TikTok are persistently confronted with such comments and, at times, perceive them as a significant source of stress.

Kellin Hanas' statements are particularly noteworthy in this regard. She cites her experience of going viral with one of her most popular videos, captioned "bop dooba doo da bwiada women in jazz babyy 🤠🎺" (Hanas 2021a). In the video, Kellin presents herself ironically as a single woman seeking a partner and displays her ability to play fast bebop licks on the trumpet, in addition to engaging in activities stereotypically associated with women, such as cooking and cleaning (see chapter 6.4.1). The video rapidly gained a considerable degree of popularity, not only on TikTok but also on Reddit and iFunny. Kellin was subjected to a multitude of comments, a considerable proportion of which targeted her physical appearance. She offers the following account of her experience:

> Oh my gosh, I've read some horrible things about myself, which I shouldn't. But it's so hard not to do when there's thousands of comments. You're, like, I'm going to read all of these and thank God most of them are good and very supportive and great. But then you get down to the bottom and you see people commenting on – oh boy, when I went viral on Reddit, I was the top post on Reddit for a day, that was just 2,000 comments. You know, maybe 500 of them were something about my playing. And then the other half of that was just people commenting on how I weighed too much, things that they would do to me sexually. Just awful, awful comments. And as a young person living in this age of social media, it's so hard not to look at what people are saying about you because you know it's there. So you want validation and you want to feel better about yourself. And even if it's bad stuff about you, you keep reading and reading and reading because it's addicting. So yeah, [...] I'm pretty sure I've read every comment at this point. It's so hard not to. [...] And so eventually, between Reddit and TikTok and iFunny it had at least six or seven million views and at least 6,000, 7,000 comments. And this was, like, the first one where people really started to take a hit at. You know, people didn't even care about the trumpet playing. It was just all about my body. [...] I've been minorly Instagram famous for a while since I was in high school in the music community. And so I would, you know, I'd get comments on my appearance all the time by, you know, older men since I was 16. But this was a different level – I have 6,000 people talking about my body. [...] So there

> have been times where I've been – I really like the video. I think it's funny. I think I did a good job with it, but I definitely have – there have been points where I've been, like, oh, I regret uploading this. I wish I never uploaded this. It doesn't make me look bad. It's just that disgusting people took it in a different way. [...] So, this was definitely one of those videos for me where I'm, like: You know, it went viral and it's a good video, but oh boy, the aftermath of it in terms of who was seeing it and talking about it was concerning and definitely really, really hard to deal with. (Kellin Hanas, interview)

One recurring argument is that musicians are often judged on the basis of their appearance in the context of TikTok because the platform reproduces and perpetuates certain stereotypes that have long been prevalent in the entertainment industry. Caity Gyorgy notes that similar issues are present on other digital media platforms and in other contexts:

> I think it's not limited to [...] TikTok. That's on Facebook, it's on Instagram as well. It's very interesting. It sucks. But to an extent, I mean, that's sort of the culture. If you look at pop stars and stuff as well. Like, the pop star isn't gonna go out on tour and not wear crazy outfits and have their hair and makeup done. I'm not gonna go to an awards show on the weekend and not have my hair done and not wear fake eyelashes. (Caity Gyorgy, interview)

Stella Cole, who has also worked in television, film, and theater, offers a similar perspective, emphasizing that in these industries, appearance is the most crucial criterion:

> Like I said, I come from a background in theater and a background in film and television. And when you're being cast in film and TV, you know that half of it, if not more than half of it, is based on the way you look. It's based on what someone looks at you and what their immediate impression is of you. Just based off your appearance, that matters so much. Like, what you're wearing matters, what kind of makeup you're wearing matters. Your hairstyle, your hair color, your weight. I wish all of that didn't matter. And it's the most frustrating part of being an actor for me, certainly. But what I'm saying is I'm used to having all of that matter. And so it's not surprising to me that on TikTok that also matters because I think there is a culture, not only in America where most of my following is, but worldwide that – when people are conventionally attractive, they're valued more to society. Their art is valued more. Most pop singers are pretty conventionally attractive, and if they're not, then they

> change parts of their face or body or hair until they do fit into the convention. (Stella Cole, interview)

It is evident that female artists have historically been subjected to considerable pressure to conform to conventional standards of beauty as stage personas. This phenomenon has also affected female jazz musicians for decades (see McGee 2009). This suggests that the practice of promoting a narrow standard of beauty in the entertainment industry is being perpetuated on TikTok, while algorithms are now influencing such processes on digital media platforms as well, potentially exacerbating the strong fixation on what is considered to be the normatively beautiful body type.

Nevertheless, the interviewees also expressed a desire to disassociate themselves from the perpetuation of these stereotypes on TikTok. In reference to a duet video with Laufey in which she is seen wearing a loose-fitting sweatshirt, Rachel Chiu stated that she initially had reservations about recording and posting a video in that particular outfit. Ultimately, however, she decided to proceed with the video, citing a certain degree of self-assurance as the motivating factor behind her decision:

> When I filmed that one with Laufey, I thought: Should I go change? Is this, you know, unprofessional? But at the same time, this is part of who I am. I love wearing this thing and, you know, it's comfortable and I'm not gonna change just to do the video. So, I just did it. There are videos where I'm not wearing makeup. Where it's just who I am. But I think that does have to do with the level of confidence as well. (Rachel Chiu, interview)

Stacey Ryan underscores her intention of moving away from conventional staging techniques in order to prioritize the musical aspect. She attributes this shift to her growing success and extensive experience on TikTok, suggesting that certain aspects of popularity, which are often extremely important to content creators on the platform, may no longer be as crucial for her:

> I tend not to think about that kind of stuff anymore, because you always do when you're starting out, but the more you go on, the more you get comfortable, obviously, and the more you just don't care as much. And I don't mean that in a bad way. You just don't really care if someone says: Oh, your hair looks gross, you know, because you're, like, yeah, it does. And? What's that going to change about the music in the way that this video sounds? (Stacey Ryan, interview)

Nevertheless, as the preceding discussions have demonstrated, it may prove challenging for musicians to deviate from these conventional staging norms. This is because the popularization logics that shape the content on the TikTok platform exert a significant influence, prompting content creators to replicate established stereotypes surrounding body standards and beauty ideals.

It is evident that the sexist stereotypes prevalent on TikTok are not the sole factor contributing to the platform's problematic content. Racist statements also abound in the comment sections. The interviewees are acutely aware of this issue, with Stella Cole emphasizing that she believes it is "easier for people who are *white* to succeed on TikTok's platform," and subsequently speaking of a "colorist or racist algorithm." Rachel Chiu has a particular insight into this phenomenon. While the majority of comments under her videos are positive, some users occasionally engage in the reproduction of racial stereotypes about Asian people. Rachel offers the following commentary on this phenomenon:

> I've talked about all the positive comments I get, but I also get a few negative comments here and there, I also get some strange comments. There's a lot of comments about *race* for sure. Like, one comment I get a lot is about me being Asian and maybe Chinese. And I think they mean it in a good way, but sometimes it comes off a little backhanded, like, they would make comments, like, oh, we need more Asians in jazz or there's not enough of that. [...] To me, it's, like, they're putting a label to it, you know what I mean? And also, I think it shows their lack of understanding and maybe visibility to other ethnicities because I know there's a lot of Asians in jazz. There's a lot of people I know who are Asians who do jazz. I watch a lot of videos of people in Japan doing jazz music. It's a whole different field there. [...] And so, I think *race* definitely has a play in it. (Rachel Chiu, interview)

Furthermore, she has observed that discussions in the comment sections frequently deviate from the subject matter of the videos themselves. The comments on one of her videos, in which she plays the guzheng, serve as an example to illustrate this:

> I've been really lucky to not have a lot of mean comments or anything like that. But usually – it's, like, with the video of me with my guzheng. I noticed that's where a lot of the comments about *race* or ethnicity comes in. Yeah, I was playing that and out of nowhere, people started talking about the Chinese government and they started talking about the different sides of it and stuff like that. [...] [T]hey're like: Oh, can you play this exact song again? But

> this time, can you wear your traditional clothes? So I think they mean it in good ways, but it does come off a little strange to me sometimes. And it does make me a little uncomfortable. (Rachel Chiu, interview)

Rachel's description demonstrates that the perpetuation of discriminatory attitudes and behaviors on TikTok is not limited to the expression of sexist stereotypes. Rather, multiple forms of discrimination often coexist and reinforce one another. This is particularly evident in the specific context of jazz on TikTok, where the analysis of the corpus reveals that the majority of individuals who achieve success in this domain are *white*. It is also important for interviewees to emphasize that individuals can identify a niche for themselves on TikTok, regardless of conventional notions of physical attractiveness. This is a reasonable assumption given the vast amount of content and the large number of content creators and users on the platform. However, it does not appear to be the case with regard to the popularity peaks of specific cultural spheres. In this regard, the results of the present study corroborate those of Kaye, Zeng, and Wikström (2022, 95–96) on the homogeneity of the most successful TikTok creators, who are predominantly young, *white*, conventionally attractive U.S. Americans. It can be reasonably inferred that, with a few exceptions, the group of individuals who are able to assert themselves in the field of popularity peaks on TikTok is primarily homogeneous.

However, based on Caity Gyorgy's observations, TikTok also occasionally serves as a constructive platform for feminist discourse. She has frequently received support from her community when her videos have been met with derogatory comments from male users. This is where TikTok differs from Facebook, according to Caity. In particular, male users on Facebook have been observed to prevent certain discussions from taking place in the first place, as Caity recalls. She cites her experiences with sexism during jam sessions and entrance exams at conservatories, where she was repeatedly labeled as less competent than her male counterparts due to her gender. She has, on occasion, disseminated and evaluated such experiences via the social networks she utilizes, and she also articulates similar criticisms in her TikTok videos. She currently prefers to communicate these issues on TikTok, where she can rely on the most robust support from female users:

> I don't post on Facebook anymore, because all of these men and people in my community that I really looked up to were just saying: Oh, this is not happening, this isn't true. And I'm, like, it is happening, you're just being an ass! So I

> think TikTok is great because you don't really get that. And if people do comment stuff like that, and are, like, this doesn't happen, then you get an army of other women in music. [...] You know, there's this great backing. There's this great support system, and I love that. (Caity Gyorgy, interview)

In light of this experience, Caity suggests that TikTok can also be used as a platform to draw attention to specific grievances, particularly sexism, within the field of jazz. Caity asserts that it is not imperative for musicians to create solemn-looking videos in order to convey such matters. Given the prevalence of comedic content on TikTok, musicians have the potential to address more profound issues with subtle allusions, provided that the audience can relate to the content:

> There's the ones about, like, you know, being a woman in the music and lots of people can relate to that too. And I also think it's good to post that because a lot of people in the comment sections on those videos are, like, I had no idea this was happening. Like, I can't believe this happened to you and I'm, like, well, it has and it has not stopping and it keeps happening. And so I guess the comedic ones are just sort of based on my experiences and I post them sort of to be funny but also just to sort of raise awareness. (Caity Gyorgy, interview)

It can thus be argued that the platform may provide musicians with the opportunity to disseminate feminist or other socially critical messages. Kellin Hanas, for instance, employs a comparable approach in her videos, presenting the role of women in jazz in a manner that is both humorous and critical. Nevertheless, the analysis of the popularity peaks in jazz on TikTok reveals that the platform primarily incites actions that perpetuate sexist stereotypes, rather than prompting a critical examination of these stereotypes.

6.8 TikTok as "Opportunity Generator"

The preceding explanations demonstrate that musicians frequently invest a considerable amount of time and energy in their TikTok careers and in professionalizing their activities on the platform. Consequently, it is evident that their presence on TikTok constitutes a significant aspect of their artistic output. Nevertheless, none of the interviewees perceived TikTok to be the primary aspect of their musical endeavors or a substantial source of revenue. Rather, for the interviewees, TikTok constitutes, to varying extents, merely a single – albeit crucial – component of their overall career trajectory, and not necessarily a comprehensive representation of their creative work. Nevertheless, all of the musicians interviewed supported the view that the platform represents an exceptionally potent instrument for rapidly and relatively effortlessly augmenting one's popularity, which is of paramount importance for the advancement of a flourishing musical career.

The ability to increase one's prominence and reach is not exclusive to TikTok. As Caity Gyorgy notes, TikTok is also an effective platform for acquiring followers across various digital platforms. She has observed this phenomenon particularly with regard to Spotify:

> I've put my Spotify link in my bio on TikTok, and as soon as I did that I started getting Spotify followers like crazy and I was, like, oh my goodness. This is something that is very, very useful to get my music across to a whole different demographic. I think my demographics on Spotify used to be, like, mostly above, like, age 35. And now, like, that 18 to 25 year old demographic has really grabbed up. (Caity Gyorgy, interview)

A certain degree of recognition on TikTok has been observed to have a beneficial impact on musicians on other platforms. Additionally, in some instances, it has resulted in the interviewees being brought to the attention of prominent media organizations that extend beyond the digital platform context. This is particularly evident in the case of Rachel Chiu, who was a contestant in the twentieth season of the talent show *American Idol*, which was broadcast on ABC in 2022. Rachel states that numerous scouts are tasked with searching for prospective contestants for the television program and utilize platforms such as TikTok. She did not submit an application to participate in the show herself but she was discovered on TikTok and subsequently invited to attend the casting event:

> They have people who are, like, scouts. And they found me on TikTok. […] I got a phone call and someone asked me: "Do you want to join *American Idol*?" And I thought it was, like, a prank call or, like, a scam because I've actually searched up how to sign up before. And it says you can't sign up unless you're American. I'm from Canada, so there's no way you could sign up. And when the person called, I was, like, no, I'm okay. And so I hung up. And then again, they called me a week later and then they were, like, would you like to join? This is real, you could search my name up, I'm a producer. I won't say her name but I searched her up and she was real. She was on IMDB and all that stuff.

> And so I was, like, okay, maybe, you know, for now they're not asking for my credit card, they're not asking for my personal information, so I'll just go with it and see how it goes. And we did some online Zoom auditions throughout the summer. There were two rounds and I remember it was very fast pace. I actually had work that day and I asked my boss. I was, like, please, can I go to my interview? And I had to drive home to my piano. Go on the Zoom meeting, and they said it would be for half an hour. So I did it, and it was half an hour, and then they're, like, wait, we need you for the next round. And so it took another hour. But I got through the online one and then they send me an email, like, congratulations, you made it. And they flew me out to Texas, which is one of their audition locations. And that's where I got to actually go and audition in front of Katy Perry, Lionel Richie and Luke Bryan, and it was super exciting. (Rachel Chiu, interview)

Rachel's story indicates that she did not anticipate these developments at the outset of her TikTok career. Nevertheless, the prospect of establishing a media presence beyond the platform is one that the interviewees are keen to pursue. Erny Nunez's experience was similar. He was featured on FOX's *Alter Ego* in 2021, at the age of seventeen, and was also head-hunted by the program's staff via TikTok:

> TikTok actually helped me get this TV show. I didn't expect it would ever. I was just making my videos one day and I got a message from Fox, saying: "Hey, I want you to be part of this TV show." And I was just, like, what? This isn't real. This is not happening. And I thought it was fake. So I was just, like, you know what? Let me just take the chance and let's see how it turns out. And it turned out to be *Alter Ego*. [...] I did love the show a lot. They use a lot of technology. I wasn't used to that. I'm used to use, like, you know, less technology when I record videos and just going from that to immediately using all sorts of things, it just it blew my mind away. So *Alter Ego* is one of those shows where you just either get it or you don't. At the moment I am going on another TV show, but I cannot say what show it is just yet. (Erny Nunez, interview)

This evidence indicates that TikTok has the potential to facilitate the growth of musicians' careers, providing them with access to more conventional mass media formats and consequently enhancing their popularity. It is evident that TikTok represents a pivotal platform for the musicians who participated in this study. Nevertheless, it seems that appearing on major television programs such as *American Idol* and *Alter Ego* is of even greater significance for the interviewees. This is suggested by Rachel and Erny's descriptions of their feelings of euphoria and excitement at having the opportunity to participate in these formats. It would appear that achieving success on TikTok, which can be considerable, is not the overarching objective for musicians. Furthermore, the allure of traditional media formats, such as TV shows, persists despite the ascendancy of digital platforms.

In the course of the interviews, the relevance of TikTok for musicians was addressed. Some of the interviewees stated that they had initially feared that their presence on the platform might even have a negative impact on their musical careers. This apprehension is largely attributed to the prevailing perception of TikTok as a platform where relatively young individuals disseminate frivolous or sentimental videos, with a paucity of serious content. Initially, the interviewees expressed concern about potential ridicule or even judgment from other jazz musicians due to their engagement with TikTok. They stressed the importance of upholding the seriousness of jazz as an art form, emphasizing that the objective of TikTok videos should not be to disparage or poke fun at jazz. Consequently, they had reservations about their credibility within the jazz community, as Kellin Hanas articulates:

> Some people in the jazz community were, like, you know, oh, this is corny. So I kind of stopped because I felt really insecure about what I was making. [...] But now I'm even more self-conscious than I was worried about. Like, what do people think about what I'm making? Do people think that I'm making fun of the art form? Do people think that I'm corny? (Kellin Hanas, interview)

In contrast, Stella Cole has first-hand experience of the fact that the majority of professional jazz musicians are not deterred by the prospect of engaging with TikTok. Rather, they recognize the potential offered by the platform, particularly in terms of expanding their reach. Nevertheless, she too initially feared that her TikTok presence would be met with rejection from the jazz community:

> When I first got onto the jazz scene in New York, I actually got really insecure about my TikTok and I was, like, oh, they're not going to think I'm a jazz singer because I'm just singing, like, "Fly Me to the Moon." Like, maybe I'm singing the wrong stuff. Or this isn't, like, truly jazz, you know? I definitely did feel insecure about that at some point, especially when I first got onto

> the scene. Because doing it on the internet is very different than doing it in person, obviously. But what I've actually found is that 95% of the musicians, singers, everybody that I meet, they're, like, wow, that's so cool that you have that platform. Musicians are always, like, oh, I'll play for one of your videos anytime. Like, I'll do a live stream with you. Like, they're very interested in it because all of them love this music and, like, any way to have thousands of people seeing it, they're down, you know? (Stella Cole, interview)

However, as Stella also notes, the TikTok platform presents only a limited aspect of her musical output. At times, she is "focusing a little less on TikTok and a little more on singing in real life." This illustrates once again that TikTok has a preference for certain types of jazz-related content over others. Consequently, musicians must align their activities on the platform with these preferences, even if their overall musical output differs from the content they create for the platform. It was similarly important for Sam Ambers to highlight that his artistic endeavors extend beyond TikTok:

> I think it's all well and good having casual fun videos on TikTok. But in terms of Spotify, you know, that's where the serious music comes into play and that's the true reflection of you as an artist, I guess. With TikTok, you're jumping through hoops, you know, you're ticking boxes. Whereas at Spotify, you want it to be as close to an accurate representation of your musical self as possible. So going forward, hopefully that's going to be the case. I'm looking to do a lot more jazz work and that will probably be more kind of commercial stuff for engagements, corporate things, weddings, whatever. I also want to have my own band that will be releasing music itself, and whether I release music through the band or independently is still to be seen. (Sam Ambers, interview)

In light of the interviewees' extensive reach on TikTok, it is pertinent to ask to what extent platform-specific activities can be monetized. Strikingly, the interviewees assert that they generate minimal to no income directly from TikTok. Nevertheless, the majority of interviews indicate that musicians can potentially derive financial benefits from their TikTok reach, albeit indirectly. Caity Gyorgy, for instance, states that the TikTok guidelines prohibit her from monetizing her videos directly. The TikTok creator fund provides financial compensation to selected content creators who meet specific criteria in terms of reach and origin. The program is open to content creators based in the United States, the United Kingdom, Germany, France, or Spain who are at least eighteen years old, have a minimum of 10,000 followers, and have generated a minimum of 100,000 views within the past thirty days. Initially, only content creators based in the United States were eligible to apply for the creator fund (TikTok n.d.; TikTok 2021a; TikTok 2021b). At the time of the interview, Canadian national Caity Gyorgy was unable to access the creator fund. However, she utilizes her popularity on TikTok to draw users to her profiles on other platforms, as previously mentioned, and to generate income. She is currently able to pay some bills based on her Spotify streams, which were indirectly facilitated by her presence on TikTok. She also receives offers for live performances through TikTok:

> I'm Canadian, I don't have access to the creator fund. [...] I don't make any money from TikTok directly. But indirectly, yes. [...] [I]ndirectly, more people [are] listening to my music [on Spotify]. They add up. I pay my rent – even more than my rent – every single month with Spotify streams. So it helped that way. I've gotten some other really great opportunities through TikTok where I've been hired to do singing work. And, I mean, it's not enough to sustain me, like, I would need one of those every single month to be able to, you know, live. But there have been opportunities that I've gotten from TikTok indirectly that have paid very well. I haven't done any sponsorship stuff before. I'd like to, like, look into getting to that. [...] For me, nothing directly from the app, only indirectly through, like, Spotify or from people, like, going to my Instagram and buying a ticket to a show or buy my album on Bandcamp. (Caity Gyorgy, interview)

Kellin Hanas has stated that she has received numerous invitations to perform live via TikTok. On occasion, these offers have provided her with the opportunity to collaborate with high-profile media personalities, which is not likely to have occurred without her substantial reach on the platform:

> I really started to get hired a lot from TikTok. Like, once I started becoming popular on TikTok, the actor Darren Criss – he was on *Glee* – reached out to me after one of my TikToks and was, like, I want you to be the trumpet player for my Christmas show. So I ended up playing with Darren Criss for Christmas because of TikTok. So I see it as a way to get my music out there because everybody's on TikTok now. [...] I mean, it's important to me. (Kellin Hanas, interview)

The second Canadian interviewee, Rachel Chiu, expressed a similar viewpoint to Caity Gyorgy regarding the potential for financial gain. She too does not receive any direct remuneration from TikTok, but instead views the platform as an "opportunity generator." Her presence on TikTok has not only resulted in her being selected to participate in *American Idol*, but has also prompted offers of collaboration from prominent corporations, some of which she has accepted:

> As a Canadian, I can't access the creator fund on TikTok, which means I can't make revenue directly from my videos. [...] I see TikTok more as a bridge to opportunities. For example, *American Idol* or different job opportunities for getting my name out there, so that in the future, if I do release music, I will have a platform to post on. I don't see it much as a revenue generator [but as an] opportunity generator. I have made a small revenue on, like, doing live streaming on TikTok, but I wouldn't say this is where I want my job to be. [...] I kind of like it that way because I feel, like, if there's monetary value placed on my videos, then it becomes a lot more, like, work for me and takes my passion away from it. [...] It's just opportunities and collaborations. I make money indirectly. For example, I did my collaboration with HP or I have done different collaborations with smaller companies. Those brands will either send me their products or I get a small pay or both. (Rachel Chiu, interview)

Collaborations such as the one referenced by Rachel with the U.S. computer and printer manufacturer HP represent a significant source of revenue for content creators on digital platforms. Content creators present a curated selection of products and frequently share discount codes from collaborating partners that offer their followers a lower price on these products. A number of the interviewees indicated that they frequently receive requests for such collaborative projects. This illustrates that jazz musicians operating in digital environments are occasionally engaged in fully commercial arrangements and, on occasion, exploit these remunerative prospects. This is, in principle, an obvious conclusion, and is a common business model in the context of digital platforms. However, the idea of anti-commercialism is, in fact, an essential aspect of representations of jazz history and aesthetic concepts within the context of jazz itself, as Scott DeVeaux (1991) elucidated in the early 1990s. The concept of commerce is viewed with skepticism in this context, and many jazz enthusiasts and performers believe that the genre should be free from the influence of contemporary market forces. A similar set of narratives is observed in a number of musical genres, and the term *commerce* continues to evoke strong emotional responses among fans and musicians alike. It is, nevertheless, a reasonable assumption that young jazz musicians are driven by the desire or the need to generate income, which may lead them to capitalize on such opportunities. From their perspective, monetizing their TikTok presence is an integral aspect of their professional development as artists. For instance, Brooklyn Stafford indicates that she began working with a manager with the objective of having a professional partner present during her collaborative endeavors:

> He helps me manage my emails if I have, like, business collaborations, right? Some people want to send you clothes, some people want commissions. And he helps me set up fees. For example, like, Beanie Babies. They reached out to me and they wanted me to do a little ad. [...] I didn't know what to charge. I don't know anything about that. But he's, like, ok, we're going to charge that and see what happens. And they did. They pulled through, and it was really good because obviously your artist manager gets a percentage of what you make, right? And so they're always going to advocate for you to give more money. So it's a win-win for everybody. (Brooklyn Stafford, interview)

Beanie Babies are soft toys produced by the U.S. company Ty, for which Brooklyn uploaded two commercials to TikTok. This example demonstrates that the musical artists' advertising partners need not have a direct connection to jazz, or even music, for these collaborations to be effective. It is similar to the collaboration between Rachel Chiu and HP. On occasion, Brooklyn notes, music-related companies inquire about potential collaborations, such as the app developer Simply, which offers applications for learning musical instruments. Brooklyn created one of her most popular videos, an adaptation of the third movement (Rondo "alla Turca") of Mozart's Piano Sonata no. 11, in response to a request from Simply (Stafford 2022):

> A lot of apps reach out to you and they want you to make a video to advertise their platform. For example, one of the videos that actually went viral was because of an idea that an app gave to me. Simply Piano, they wanted me to record [a] Mozart piece. And they said, maybe you can do a pop aspect to it, but I wanted to do a Latin remix to it. [...] I posted that on my Instagram and TikTok and those both went off, but that was actually for an ad. And so I gave the ad to them and they posted it and whatever. And that's the origins of that. So yes, in long story, yes, you can make money off of TikTok from apps and people reaching out to you for commissions. (Brooklyn Stafford, interview)

Nevertheless, the responses of the surveyed musicians on the subject of collaboration differ widely. For some, collaboration with companies is an unfeasible proposition, as Erny Nunez affirms:

> I've been offered, you know, some collaborations to promote, like, a product, but I always declined because I don't want to gain money off of TikTok. That's not what I made the account for, you know. Yes, I've got a lot of, like, companies saying, hey, I want you to promote this product. But that's not really what I find fun. I find it better, you know, if I just sing. I just like doing what I do. It's, like, a hobby. (Erny Nunez, interview)

Sam Ambers, on the other hand, says that he is amenable to collaborations, provided that they align with his TikTok profile. He asserts that the crucial consideration is to only engage in collaborations that he and TikTok users can understand. A potential collaboration with Samsung, for instance, would be a viable option. A few weeks after the interview, he did in fact post a commercial video for Samsung (Ambers 2022b). Sam characterizes his stance on collaborations as follows:

> I've got of a lot of friends that I've known through TikTok that do a lot of personality-based videos. So it's not music or anything. It's maybe a bit of comedy and, you know, they can get brand deals very easily. And let me tell you, they're making a lot of money. Whereas, you know, with my account, you want to keep it authentic to a certain degree. So that means that you can't sell out to brand deals and things like that because what is that? You know, that is almost a bit of a stain on your kind of artistic portfolio or the way that you're viewed, basically. I think I do need to be a bit smarter. I do need to be a bit more clever with TikTok because there are really good opportunities and there are brand deals that are viable and that I'd absolutely love to do. I think the obvious one is Samsung. I think that absolutely makes sense. I have absolutely no problem with that. And if there was a brand deal that I could deliver in an entertaining, engaging, artistic way, there's nothing wrong with that whatsoever. But, you know, I think the goal for me is to become a musician, not an online personality. So that's kind of where I draw the line. (Sam Ambers, interview)

While attitudes toward collaborative opportunities and sources of income vary, a certain fundamental proficiency in self-marketing appears to be a universal necessity in the contemporary music industry. Stella Cole, for instance, says that musicians must initially establish a certain degree of reach within the platform context to capture the attention of record labels and management. Subsequently, they must possess the capacity to effectively delegate such marketing responsibilities when necessary:

> It's so interesting that artists are now expected to – I mean, artists have always had to market themselves, but now it's not, like, you sign to a label and then they market you and they're, like, okay, this is what your brand should be. This is how we're going to get you exposed to people. It's, like, you do the work of yourself to the world. And then a label will find you, like, Stacey [Ryan] signed with [Island Records]. They wouldn't have wanted to sign her until her video got eight million views. (Stella Cole, interview)

TikTok appears to provide optimal conditions for the form of self-marketing described by Stella. The interviewees concurred that TikTok is a relatively low-threshold avenue for reaching new listeners, and that this can be a pivotal factor in advancing a musical career beyond the platform. Sam Ambers offers a perspective on this: "Effectively, it has to be used as a platform to ultimately assist a musical career, which is kind of what I want to develop." For Caity Gyorgy, the primary benefit of TikTok is its capacity to disseminate content to a vast audience in a relatively short period of time, extending even beyond the confines of the platform. This represents a novel experience for her, and a crucial distinction from other social media platforms such as Facebook and Instagram. Caity elaborates:

> I've always been super interested in, like, social media marketing. But I had mainly done that through Instagram and Facebook and then along comes TikTok, and it just made a world of difference. [...] I'm very grateful for the app. [...] [W]hen [one of my "Cardiologist" videos] went viral, I gained 2,000 followers on Spotify. [...] When I started TikTok, I had around like 2,000, 2,500 Spotify followers, and now I'm up to over 9,500 Spotify followers and I credit that entirely to TikTok. It's been pretty fantastic. It's really nice to get the music around to different people. [...] Instagram is very limited with the amount of reach you can get to other people and, like, I have a transcription account on Instagram, that's taken me years to grow and I haven't posted on it in a while because I've been so busy. But even now, like, I would post and I'm, like, oh, I have a new single! Nobody cares about that on Instagram. But on TikTok – it's pretty phenomenal. [...] If they hear a song they like, if they hear lyrics that they resonate to, if there's a beat that they find interesting or a

> melody that catches their ear, they like it and they interact with it. And that's something that I've never experienced with apps like Instagram or Facebook before. I think the potential of reaching new people is insane. (Caity Gyorgy, interview)

Erny Nunez highlights that TikTok facilitates the "biggest amount of traction," a notable distinction from other platforms: "I started on YouTube. I didn't like it. I didn't really like YouTube. It was very slow. It was very random based. [...] Instagram same. Anything else that I used, SoundCloud as well. So I would say TikTok is extremely important."

The interviewees frequently compare the platform to Instagram. It is noteworthy that all of the musicians interviewed are active on both platforms and that, without exception, they had significantly fewer followers on Instagram than on TikTok at the time of the interviews. They attribute this to the different functional logics of the platforms, with TikTok being characterized by a "virality-centered platform logic" (Zeng and Kaye 2022, 80). The prevailing view is that viral moments are uncommon on Instagram because the platform is designed to facilitate communication with users who have actively chosen to follow a particular profile. This is significantly less the case on TikTok, where the platform's For You page logic enables content creators to reach users with whom they have not yet established a connection. Kellin Hanas provides the following explanation:

> You can go viral on TikTok, but it's very rare to go viral on Instagram because that's just not how the platform works. [On] Instagram [...], you connect with the people that you already know. And sometimes you get lucky with your videos and they reach a wider audience. But it's pretty much who already knows you is going to follow you. [On] TikTok, [...] you upload a video. That video goes out to an audience of random people that the algorithm selects and it keeps going up and up and up to more random people. So it's just natural for me to upload a video on TikTok that goes viral. Then I've got 50,000 random people start following me and only 25% of them will look and see, oh, here's her Instagram and then follow me on Instagram. [...] It's just easier to gain people on TikTok. (Kellin Hanas, interview)

Sam Ambers expressed a similar sentiment, pointing out that he probably only gained a few followers on Instagram because those people were already familiar with his TikTok account:

> I feel like it's quite easy to get views and develop a following on [TikTok]. [O]n Instagram, [...], unless there was an exceptional amount of attention or traction to the video, it would only get shown to my followers. So there's such little room for growth. Any growth on Instagram tended to be from external sources. So the only reason why I've got a few thousand followers on Instagram is because my Instagram is attached to my TikTok. It trickles down followers. [...] I mean, the case in point is that I had zero followers on TikTok when I posted my first video and then overnight I had 10,000 followers and, like, two million views. You know, that doesn't happen on Instagram. (Sam Ambers, interview)

It is noteworthy that other platforms apart from Instagram are only briefly referenced in the interviews. While some of the interviewees do mention Facebook, they seldom discuss other platforms such as YouTube, SoundCloud, Bandcamp, Twitch, or X (formerly Twitter). It is evident that the interviewees' age is a contributing factor. Individuals under the age of twenty-five are currently more inclined to utilize platforms like TikTok and Instagram. Conversely, as previously mentioned, Caity Gyorgy and Erny Nunez have provided insights into their experiences with other platforms, which are at times a cause for concern. It is similarly conceivable that the particular and relatively homogeneous content for which the interviewees became known on TikTok may not be equally popular on all platforms due to the presence of different functional logics and target groups.

6.9 TikTok's Potential Influence on Jazz Culture

It has been demonstrated that certain types of jazz-related content are more visible on TikTok than others. The musicians interviewed are aware of this phenomenon and, as a result, develop specific strategies and occasionally produce videos specifically for the platform with a high degree of precision and according to clearly defined criteria. In light of the considerable influence that musicians wield on TikTok, it is worth examining the extent to which the platform is shaping the public perception of jazz. The interviewees concur that TikTok presents a valuable opportunity to introduce jazz to new audiences, particularly younger individuals. In this context, the musician Laufey is referenced on numerous occasions. She is regarded as one of the most prominent jazz musicians on TikTok and is described in various journalistic articles as teaching jazz to the "TikTok generation" (Dellisanti 2022) or "Gen Z" (Katel 2022) and as "bringing jazz to a new generation" (Williams 2022). Stella Cole also highlights Laufey's capacity to expand the reach of jazz to a broader audience, emphasizing that such advancements are inconceivable without the influence of TikTok.

> [Laufey] was one of the first creators that I followed and that followed me back. And now she's obviously, like, blowing up and doing such amazing work. And she's really been able to bring jazz to the pop charts, which is so incredible. And she started on TikTok, she would not have the same followers if not for TikTok. (Stella Cole, interview)

Due to the viral nature of TikTok, a video that gains traction on the platform can be viewed by a vast number of individuals in a relatively short period of time. This unique ability to reach a vast audience in a short span of time is not available in traditional media channels or other platforms. This indicates that the dissemination of jazz-specific content to young people who may not have had a prior affinity for this musical genre is a viable avenue. The functional logic of TikTok presents a promising opportunity for the propagation of jazz. Nevertheless, some of the interviewees initially questioned whether there was a demand for jazz on TikTok and whether the platform was an appropriate space for this genre. However, according to Caity Gyorgy, the TikTok audience appears to be receptive to a diverse range of music-related content, largely transcending genre boundaries:

> I honestly thought, like – do people these days, children essentially – because a lot of these people on the app are under 18 – do these kids want to hear this music? Like, do they even care about this? And it turns out that they actually do. It's not really a genre difference. If they hear a song they like, if they hear lyrics that they resonate to, if there's a beat that they find interesting or a melody that catches their ear, they like it and they interact with it. (Caity Gyorgy, interview)

Kellin Hanas posits that the rejection of jazz by younger demographics is not a direct rejection of the genre itself, but rather a consequence of the lack of exposure to it. She further suggests that the challenge lies in the introduction of jazz to a diverse range of listeners: "[T]hey're not interested in jazz themselves, but when jazz is brought to them, they're, like, ah, jazz!" Caity Gyorgy also emphasizes that jazz does not often "get to the ears of the average consumer," and that TikTok offers the opportunity to change that. As the musicians observe developments in the comments to their videos, they formulate concrete ideas of how to communicate the phenomenon of jazz's appeal to younger demographics. They examine the reasons behind the popularity of specific jazz styles and older jazz songs on TikTok. Stacey Ryan suggests that, based on her experience, a significant proportion of TikTok users may lack prior familiarity with jazz and its various forms before encountering it on the platform, which may potentially lead to an affinity for it. The reasons for jazz's popularity on TikTok are multifaceted, according to Stacey:

> I think it's a mix of a couple of things. I think part of it is that no one ever really listens to that. So they never got introduced to it anyway. Like, you know, their parents never listen to that kind of stuff. They never listen to it, they didn't go to music school, so they didn't have to listen to that kind of music. So when they see it on TikTok for the first time, a lot of them like it without even knowing that they do because they're taught or their friends tell them that jazz sucks and it's bad. And if you have that opinion, that's fine. But a lot of them don't even know because they've never heard it. So when they see this young person doing old-style music, a lot of them are intrigued. Some of them are, like, this sounds like actual shit and that's fine. But most of them are, like, wow, I've never heard this kind of music before. Or, oh wow, I like jazz now. I didn't even know. So I feel, like, part of the people just not knowing is what makes it really popular. [...] I mean, for all kinds of music, you know, it takes, like, not all the time, but, like, theory knowledge and practice and everything. But because jazz is more complicated, just in theory to understand, I feel, like, people are also kind of impressed when they see, like, this teenager looking person is executing, like, maybe a more difficult song or, like, with very jazzy chords. For some people that's too much. They're, like, it's too much for me. I like it and that's fine. Like, I've gotten my fair share of comments of that, but a lot of them are, like, wow, I've never heard this before. And I love, like, the super intricate chords and with the notes you're singing on top of them. So I feel like those are a couple of things that definitely make jazz more palatable, I guess, if that makes sense, because they're being fed to it or they're being fed it on a very pop, very mainstream platform. So I feel like they're like, oh, well, TikTok is showing me this, then I probably like it. (Stacey Ryan, interview)

From Stacey's perspective, the prevalence of jazz on TikTok may contribute to a reduction in the hostility directed towards this musical genre. This is due to the fact that a particular and relatively accessible aspect of jazz is currently a dominant feature of the platform. It is therefore important to ensure that jazz is brought to the attention of its potential audience, as Stella Cole also emphasizes: "I feel, like, there are just a lot of people in the world and on the internet who don't really know that jazz is still such a big thing. [...] [P]eople aren't exposed to this music," Stella assumes, and TikTok could help to "move [jazz] forward to different audiences that wouldn't have otherwise been exposed to it." Such arguments are sometimes accompanied by the belief that TikTok might even help preserve jazz. Erny Nunez speculates that by exposing more young people to jazz, the songs from the Great American Songbook that are so popular on TikTok will become part of the consciousness of future generations:

> I think this type of music, compared to what, you know, what's listened to now is very wholesome, I would say. It not sounds old. This type of music – it's just made to sort of dance. That's what I would describe this music as. It's just made to enjoy, have fun, just being happy. And so I guess seeing it blow up on TikTok gives a lot of traction to the type of genre, which I'm very happy for. And I hope it inspires many, many other young musicians to continue. (Erny Nunez, interview)

The musicians interviewed thus attest to the platform's potential to contribute productively to the preservation and dissemination of jazz, perceiving this as one of TikTok's key strengths. Kellin Hanas emphasizes that it is necessary to "create jazz fans," and asserts that TikTok offers an optimal platform for introducing bebop to a broader audience:

> I see it as a way to get the general population to become more exposed to jazz music, which is really important, because if we want this music to survive and continue to grow, we need to gain more traction for it. [...] And if this is the way to expose them to bebop and other styles of music that people aren't really listening to, then, great, let's do that. (Kellin Hanas, interview)

In order to facilitate such developments, Kellin maintains that it is also essential to refrain from acting as a gatekeeper in the jazz context. Instead, it is crucial to ensure that all individuals have access to this musical genre, including those who may be more inclined to consider the commercially oriented repertoire as falling outside the realm of art music. Rachel Chiu views it as her mission to exemplify this open-minded approach and to make jazz accessible to as many people as possible, particularly younger demographics: "There's a pretty big group of people who actually still appreciate jazz which I love. And that's something I'm trying to do. To keep that going, and find ways to incorporate it into my generation, and, you know, jazz isn't for old people, jazz is for everyone."

6.10 Summary of Interview Results

What facets of jazz are most visible on TikTok and why? Who is reaching a particularly large number of users on the platform with what jazz-related content? What do musicians do to be as visible as possible on the platform? What, then, is the image of jazz that emerges on TikTok? Given that the interviewees represent some of the most popular figures in the defined research field of jazz on TikTok, the results of this study are limited in their ability to provide generalizable insights into the practices of musicians in other contexts. However, the statements of the individual musicians on certain topics are occasionally strikingly similar, thus enabling the point of empirical saturation to be quickly reached during the course of the content analysis. This suggests that the findings can be regarded as representative, at least with respect to the selected research area (Strübing 2014, 32). The interview results demonstrate that jazz performances on TikTok are distinguished by particular socio-technical interaction processes between the platform, the musicians, and the TikTok users. These processes are contingent upon the algorithmic moderation of the TikTok platform, the format of the short-form video, the features and interface of the platform, the musicians' interpretations of the platform's logics, their claims to success, and the reactions of the users.

The statements of the musicians interviewed for this study can be related to the central concepts of current research on digital platforms – algorithmic culture, platform affordances, algorithmic imaginaries, and platform vernaculars – introduced in chapter 3.5. It is evident that a distinct algorithmic culture has emerged on TikTok in relation to jazz. This can be defined as a digital space in which platform algorithms and cultural workers interact closely. These interaction processes result in the formation of socio-technical constellations that could not arise in this form without the cooperation of human and non-human actors. It can be argued that the aforementioned phenomena are made possible by the typical platform mechanisms, in particular the processes of datafication and algorithmization. The specific media representation of jazz on TikTok in its current form can thus be seen as a consequence of the informatic processes of data collection and data processing.

The results of the interview indicate that the platform affords specific courses of action to which the musicians respond. Fundamentally, the musicians must find their own way of working with the prescribed format of the short-form video and become acquainted with the core functions of the platform. Furthermore, the interviewees are all concerned with the question of which content on TikTok is most likely to achieve the widest possible reach and how they can meet these demands. In this way, the platform's affordances have a significant impact on the musicians' creative work, although they do not necessarily determine it.

Moreover, the platform is conducive to the development of algorithmic imaginaries, as evidenced by the interview results. Content creators are unable to ascertain the precise algorithmic functionality of TikTok, which introduces an element of uncertainty as to which production strategies are the most effective for achieving success on the platform. However, they develop well-defined concepts of potential avenues for success based on their experience. These concepts then exert a considerable influence on strategic video production decisions. It is crucial to acknowledge that the musicians' creative concepts are not solely derived from algorithmic processes; they are also influenced by the potential responses of TikTok users. Content creators' assumptions about the behavior of human and non-human actors are not always distinctly delineated. They frequently allude to broader activity on the platform.

As a result, a homogeneous idea of jazz or, in other words, a set of jazz-specific platform vernaculars emerges on TikTok, at least in the field of popularity peaks examined in this study. The influence of these vernaculars is such that they can, in turn, influence other content creators who also aim to succeed on TikTok, or who simply follow well-known role models. This is due to the fact that they are represented by musicians with a very wide reach. In general, it can be observed that the functional logic of TikTok tends to promote the homogenization of jazz-related content. The following section provides an overview of the various aspects of jazz that are particularly prevalent on TikTok and examines the extent to which this representation of jazz is shaped by socio-technical interactions between the platform and the musicians.

It can be observed that the majority of musicians interviewed have received some form of musical education, often spanning many years of singing and/or instrumental lessons. Additionally, many are enrolled in or have already completed study programs with a jazz focus. However, it is notable that a postsecondary education in music is not a prerequisite for success on TikTok in the field of jazz, with the majority of interviewees exhibiting a diverse range of musical styles and approaches. Self-taught musicians and singers with professional training in the musical field are also active on TikTok. While there are always exceptions to the rule, jazz is generally regarded as an academic musical practice. The study of jazz is offered at a number of conservatories around the world. Jazz musicians who have established themselves in the professional music business have typically obtained an academic degree. It is evident that a considerable number of amateurs perform in big bands, for example. However, in the context of jazz, it is primarily those musicians who have received academic training who are regarded as the epitome of this musical culture today. Such criteria are apparently of lesser importance on TikTok. The musicians interviewed demonstrate a striking openness to incorporating elements of other musical genres, extending beyond the traditional boundaries of jazz. This openness to diverse forms of popular music is regarded as a crucial factor in the continued evolution of jazz. The interviewees demonstrated a lack of adherence to rigid genre boundaries or strict distinctions between jazz and popular music. This is also reflected in the musical role models they mentioned. The discussion includes musicians such as Frank Sinatra, Judy Garland, and Michael Bublé, who exemplify a musical style that bridges the gap between jazz and popular music. Consequently, they are often overlooked or, at best, only briefly mentioned in established jazz histories (see McDonald 2006; Frith 2007; Burkhart 2024).

Certain facets of jazz, which are sometimes not even recognized as jazz, depending on the geographical and institutional context, are perceived as the most popular in the context of TikTok. For example, while the interviewees unanimously identify Frank Sinatra as one of the most popular and influential jazz musicians of all time, my own experiences of lectures and courses on jazz on TikTok in German-speaking countries paint a different, sometimes even contradictory picture. In these academic contexts, there is a tendency to articulate precise and well-defined concepts of the defining characteristics of jazz, the expected and permissible actions of jazz musicians, and the precise boundaries between jazz and popular music. The popular jazz performances on TikTok are, at best, considered *pseudo-jazz*, as one participant put it at an academic conference where selected results from this study were presented. The image of jazz that emerges on TikTok and the jazz-specific platform vernaculars on TikTok therefore deviate considerably from the prevailing perceptions and representations of jazz beyond the platform. This shows that, depending on the context, there are several possible answers to the question of what specifically characterizes jazz as a musical form. For the interviewees – and most likely for many of the TikTok users who watch their videos – it seems natural to view at least certain facets of jazz as part of popular music culture. Nevertheless, within the academic jazz community, it remains a common practice to categorize jazz as art music and to differentiate it from popular music.

The results of the interviews indicate that the optimal utilization of the platform necessitates the implementation of professionalization processes that are tailored to the specific characteristics of the platform. Musicians engage in the professionalization of their TikTok-related activities by continuously observing the underlying logics of the platform and developing strategies for success that align with these observations. It is perhaps unsurprising that traditional notions of professionalism are being challenged in the context of TikTok. This is largely due to the fact that the platform is not always taken seriously in the wider public perception and is discredited as a medium of expression for frivolous content, especially content created by young adults. Nevertheless, particular forms of professionalization are emerging on platforms like TikTok, which are tailored to the specific requirements of the respective platforms. These forms of professionalization are, in essence, not significantly different from professionalization strategies that have long been common practice in the music industry. For example, YouTube was launched in 2005 with the goal of creating a platform for individuals to share their own content, regardless of the need to adhere to traditional professional production standards. The slogan was *broadcast yourself*, and the content in question was often recorded by users themselves with relatively basic digital or mobile phone cameras. At the time, there was still a relatively clear distinction between user-generated content (UGC) and professionally produced media products. However, the boundaries between the two are now blurred, and new roles have emerged that operate at the intersection of UGC and professional production. One illustrative example is that of YouTubers, namely content creators who do not necessarily collaborate with production companies. However, they have professionalized their YouTube-related activities, for instance by using high-quality cameras and microphones and sophisticated post-production processes (Jost 2017, 55–56). Similar processes can be observed on TikTok. The interviewees indicated that they frequently produce multiple versions of a single video, with some producing up to fifty versions, in order to achieve the optimal quality. The alleged spontaneity observed in many TikTok videos is therefore often an illusion. The success strategies or imaginaries that are developed over time influence a number of factors, including the choice of which technical equipment is used to record videos. While it is possible to simply shoot videos with a smartphone and upload them without any further editing, the production process is sometimes much more complex. It is frequently assumed that both image and sound quality must meet certain quality standards, which has led to investments in technical equipment (such as tripods and ring lights) and the use of professional microphones and specialized audio software for audio tracks, rather than simply using a smartphone. Conversely, the videos that garner the most attention are often those that do not initially meet conventional quality standards, leading to the assumption that they are particularly relatable for a significant number of TikTok users. Overall, it can be observed that the interviewees draw upon their experiences on the platform in unique ways when producing their videos, thereby contributing to the evolution of a distinctive representation of jazz on TikTok.

The prevailing view is that success on TikTok necessitates sustained effort and attention to detail. It is essential to interact with the platform and its functional logic over an extended period, to upload videos on a regular basis, and to disregard initial setbacks. This approach is the most effective means of achieving long-term success, which is primarily gauged by visibility – or the number of views. It is therefore advisable that videos are created with the objective of generating as many views as possible, even if this results in a certain degree of detachment between the viewer and the content. In this context, views represent the primary currency on TikTok. This is also the primary motivation for developing algorithmic imaginaries, as the ideas generated about how TikTok functions are largely related to the question of what methods can be used to generate the highest possible number of views. Notwithstanding the extensive experience of the interviewees, individual success on TikTok can only be planned to a certain extent, even for those musicians with a considerable reach who were interviewed as part of this study. The rationale behind algorithmic prioritization remains opaque, which is why the platform has the effect of encouraging the development of speculative strategies. Failure to succeed is perceived as frustrating by the interviewees. Concurrently, failures are occasionally regarded as a catalyst for amplifying reach through the creation of new videos. This can potentially result in a form of addiction, which is arguably one of the distinctive features of TikTok and a pivotal factor in its success. As the rationale behind the platform's functionality remains opaque, the influence of routine behavior is constrained. Content creators are persistently prompted to attempt a second attempt in order to achieve viral success.

This illustrates that speculation regarding algorithmic functional logics can be regarded as a fundamental aspect of cultural production on TikTok. The platform deliberately encourages this phenomenon through its technologically mediated prompts to action. As part of the content analysis, a multitude of topics were identified that are linked to algorithmic imaginaries and encompass statements pertaining to concrete strategies for self-presentation and musical production. The crux of the matter is that all the interviewees develop particular assumptions about the TikTok algorithm, yet they do not address the algorithm from a computer science perspective. Instead, the emphasis is on diverse methods of video production that the interviewees anticipate will afford them an advantage in the algorithmically moderated competition for visibility. As a result, the concepts that interviewees develop regarding *the algorithm* remain vague and abstract. Furthermore, these concepts often include assumptions about potential behaviors of TikTok users. The opacity of TikTok's precise functional logic prevents content creators from understanding the extent to which algorithmic prioritization or user reactions – or both in combination – are responsible for a video's success. It is often unclear from interviewees' statements whether they are referring to the platform's algorithmic moderation, user preferences, or both.

Beyond the question of the extent to which algorithms and user reactions determine visibility, the musicians interviewed identified specific patterns of video production that they believed made success more likely over time and through ongoing observation of what happens on the platform. For instance, the interviewees consistently stressed the perceived necessity of tailoring their performances to specific TikTok conventions, while simultaneously striving to differentiate themselves from their peers within the confines of what might be deemed effective on TikTok. It is of the utmost importance to differentiate themselves from the plethora of content on the platform in order to capture users' attention and to prioritize the inclusion of details that prompt users to not swipe a video, but to engage with it meaningfully, such as by leaving a comment or expressing approval through a like. This evidence clearly illustrates the aforementioned tendency to consider not only the algorithm itself, but also the users within the context of platform-related concepts. The central assumption is that TikTok users have relatively short attention spans. This is attributed to the functional logic of the platform, particularly the format of short-form videos and the principle of the For You page. Widespread hypotheses on a platform-related attention economy in the field of cultural production are at least implicitly referenced in this context (Leveillé Gauvin 2018). Furthermore, it can be surmised that the imaginaries of the musicians interviewed are not solely based on their own experiences in the TikTok context, but that external sources of information or widespread theories about algorithmic prioritization on digital platforms or algorithmic gossip (Bishop 2019) may also play a role. For example, Sam Ambers suggests that the initial success of his first viral TikTok video can be attributed to its being inadvertently displayed to a select number of users on their For You pages, who then proceeded to engage with the video. Sam assumes that the video was disseminated to an ever-increasing number of users, ultimately reaching millions of views despite the fact that the creator was not yet a well-known figure, as a consequence of user engagement. The hypothesis that TikTok initially distributes individual videos to a limited number of selected For You pages to assess their potential for virality on the basis of initial reactions is a widely held theory that can be attributed to algorithmic gossip. Sam Ambers' assessment is thus not based on specific knowledge, but rather on his examination of the functional logic of TikTok at a more abstract level, which also encompasses the reception of external sources.

Similarly, the interview passages indicate that the musicians believe that any form of interaction with users on TikTok can potentially contribute to their success on the platform. This means, for instance, responding to user comments. However, as indicated by the interviewees, it is equally important for them to encourage as many user reactions as possible with their videos. This can be achieved by motivating them to comment, like, share, and so forth. In order to gauge users' opinions, questions are posed in a variety of formats, including video descriptions and text layers that can be placed directly in the video frame. These can then be referenced in the comments section. These assumptions are similarly prevalent among content creators across platforms (Cotter 2019; Cotter 2022). It would appear that the specific platform to which the imaginaries refer in detail is not a definitive factor. It is evident that in recent years, discourses on the functional logic of digital platforms have emerged across various platforms. In this regard, cultural workers do not see individual platforms such as TikTok as standalone entities. Instead, they interpret them within the context of cross-platform imaginaries. It is noteworthy that, despite demonstrating a discerning attention to the nuances of TikTok, the musicians interviewed seldom direct their attention towards the platform's defining characteristics, particularly the duet, use this sound, and stitch features. For example, they may review a significant number of comments to gain a deeper understanding of users' interests. Alternatively, as illustrated by the case of Rachel Chiu, they may create an anonymous second account to improve how they use hashtags. However, it is notable that comment sections and hashtags are also key features of similar platforms, including Instagram, YouTube, and Facebook. In contrast, the interviewees use the duet feature on TikTok relatively rarely. Although they consider this feature beneficial in terms of expanding their reach, they do not prioritize it. This indicates that cross-platform algorithmic imaginaries have become entrenched, at times influencing actions to a greater extent than the specific features of individual platforms.

Nevertheless, decisions about which jazz repertoire to showcase on TikTok are significantly shaped by specific assumptions about the functional logic of TikTok. In particular, the interviewees identified considerations of what jazz-specific content can best be presented within the short-form video format, and which repertoire or musical expressions are most likely to be preferred by TikTok users, as key issues. These considerations are informed by the logic of the For You page. The results of the interviews indicate that, from a musical perspective, there are homogenization processes occurring on TikTok, which in turn strongly depend on the platform-specific concepts of the musicians.

In this context, the focus on comparatively old jazz standards that can usually be assigned to the Great American Songbook repertoire is particularly noteworthy. The fundamental premise of the interviewees is that these songs are popular on TikTok due to their initial widespread recognition. This assertion can be substantiated, at least within the context of North America. Concurrently, the interviewees indicated that it is crucial to infuse these widely known compositions with a distinctive personal interpretation in order to differentiate themselves from the vast array of content on TikTok. The interviewees believe that the ability to set themselves apart from other creators on TikTok while simultaneously aligning with the platform's distinctive norms is a pivotal factor in attaining success in the TikTok domain. This perception significantly shapes the selection of content. The interviewees' descriptions occasionally appear to be objective accounts. For example, some musicians believe that particular jazz repertoires and specific interpretive strategies are conducive to success on TikTok. Nevertheless, the strategies of repertoire selection must necessarily remain speculative. In this context, it becomes particularly evident that the musicians' imaginaries are only partially related to TikTok's algorithmic system, and that the potential reactions of users form another part of the basis of their speculations.

Most of the interviewees assert that TikTok does not provide an appropriate platform for displaying technical proficiency and musical spontaneity, despite these qualities being regarded as fundamental in the context of jazz. The emphasis on relatively brief videos precludes the presentation of more extended improvisations, which musicians assume the TikTok audience is not inclined to appreciate. The prevailing view was that original compositions were unlikely to gain significant traction on the platform. The affordances of the platform and the imaginaries of the musicians therefore exert a strong influence on the musical content that is *not* performed on TikTok in a jazz context. In contrast, humorous content appears to be particularly well-suited to jazz on TikTok, as evidenced by the videos of Kellin Hanas, Sam Ambers, and Caity Gyorgy. In this context, musicians can utilize established platform conventions, as musical content on TikTok is occasionally employed as an auditory element to help set up punchlines in humorous videos. It is notable that humorous content is pervasive on TikTok (Abidin and Kaye 2021, 58). In this context, however, the interviewees occasionally highlight that their jazz performances on TikTok are frequently distinct from their musical activities outside the platform. For instance, the style of a punchline-based short-form video may not readily translate to the context of a live performance in a jazz club. Consequently, the jazz-specific vernaculars that emerge on TikTok may deviate considerably from those observed in conventional jazz performances outside the platform.

As the popularity peaks in the field under study are largely represented by North Americans, the interview statements on the musical repertoire exhibit a certain North American-centricity, which the interviewees are acutely aware of. For example, respondents indicated that songs from the Great American Songbook constituted a fundamental component of their musical education during their formative years and were a customary feature of their Christmas celebrations. It is postulated that the association of these songs with Christmas evokes positive sentiments for a considerable proportion of TikTok users. It is also noteworthy that the choice of repertoire can be influenced by various popular media sources beyond the platform. For example, some of the jazz standards that have become popular on TikTok are featured in the soundtracks of popular television shows, films, and computer games. For example, Stella Cole is aware that some users are familiar with the song "Fly Me to the Moon" from the Netflix series *Squid Game* and do not necessarily recognize it as a jazz standard, as illustrated by the comments on her videos. It becomes evident that some of the content that gains popularity on TikTok is disseminated by traditional mass media outlets prior to its proliferation on the platform, thereby influencing the representation of jazz on the platform. Therefore, negotiations of media visibility may occur outside of TikTok prior to the algorithmic categorization procedures initiated by the platform and the user-driven promotion of specific content, and subsequently influence the platform's dynamics. This illustrates that TikTok is not an isolated entity, but rather subject to the influence of traditional mass media.

Moreover, the interview results indicate that discourses on social categories such as gender and *race* also impact the representation of jazz on TikTok. For instance, common gender stereotypes, body norms, and normative beauty ideals exert a considerable influence on the popularity of jazz on the platform. The interviewees' experiences suggest that certain body types and modes of presentation are more conducive to success on TikTok, prompting them to adapt their behavior in accordance with these perceived norms. The results of the corpus analysis conducted for this study indicate that physical characteristics are a significant factor in determining who is afforded the opportunity to become highly visible in the jazz context on TikTok. This is particularly evident in the case of *white* female North Americans, the majority of whom are younger than twenty-five and conform to prevailing beauty norms. Musicians who meet the specified criteria may then use their bodies in particular ways and customize their videos in a way that enhances their visibility.

The musicians interviewed were forthcoming and critical about this issue. It is evident that high visibility on TikTok is less likely if certain beauty practices, such as wearing makeup and certain clothing, are not followed. Furthermore, the interviewees have identified potential patterns of success in this regard. Despite their fundamentally critical stance, they adapt their strategies for video production in accordance with these patterns. Once more, it is not yet clear who or what is responsible for the popularity of the videos in question. Are they the result of the TikTok algorithm, the users themselves, or their interactions? Rachel Chiu's experience with a live stream that reached a significantly larger audience after she changed her outfit and makeup, for example, suggests that appearance plays a significant role in the algorithmic filtering process. This is due to the fact that even a live stream must initially be displayed to TikTok users on their For You pages, which are based exclusively on algorithmic selection processes. The extent to which user reactions contribute to a video being shown on an even greater number of For You pages, and the extent to which the musicians' appearance plays a role in this, cannot be definitively determined. It is evident that there is a complex interplay between algorithmic prioritization, user preferences, and musicians' actions based on platform-specific perceptions. This interplay contributes to the homogenization of musical styles and practices. Although some musicians may choose to post videos of themselves without any makeup, they nonetheless feel compelled to conform to certain standards of beauty. However, this phenomenon is more prevalent among female interviewees than among male musicians, who tend to attach less significance to this issue based on their personal experiences. Although the interviewees do discuss these issues, they occasionally assert that this issue is not unique to TikTok. The argument put forth is that certain conventions of representation based on normative notions of beauty that have long been commonplace in music and media-related professional fields (and beyond) have been transferred to the platform context. This is an irrefutable fact. Consequently, the most significant development in this regard is the involvement of algorithms in the competition for visibility on platforms such as TikTok. Nevertheless, the precise impact and homogenizing potential of these non-human actors remain opaque.

Notwithstanding the aforementioned problematic aspects, the majority of interviewees maintain a predominantly positive relationship with TikTok. However, as articulated by Rachel Chiu, they perceive the platform as an opportunity generator rather than as their primary medium for self-expression. In the course of the interviews, several of the musicians indicated that they had benefited considerably from their engagement with the TikTok platform. They observed that it had enabled them to enhance their visibility and reputation. As a consequence of their success on TikTok, some of the musicians had been presented with the opportunity to participate in high-profile television programs, including *American Idol*. This shows that traditional mass media formats retain significant value for interviewees, despite the proliferation of digital platforms. In some instances, the interviews suggest that the prestige associated with appearing in a major TV production may outweigh the influence of having a large social media following on TikTok. For the interviewees, TikTok is not a primary source of income; rather, it is a platform utilized to attract listeners and draw attention to their own Spotify profiles or upcoming concerts, for example. However, in some interviews, it becomes evident that remunerative opportunities can also arise directly on the platform, for example through collaborations. The monetization regulations on TikTok are subject to frequent alteration, which may account for the discrepancies between the interviewees' views of the present circumstances. Collectively, the prevailing opinion is that TikTok's primary objective is to facilitate the rapid and extensive dissemination of content. This, according to the interviewees, represents the primary difference between TikTok and other platforms such as Instagram, Facebook, and YouTube.

The interviewees also view the potential to disseminate their own content to a vast audience with minimal effort and in a relatively short space of time as a positive development for the future of jazz. In conclusion, the platform provides a valuable opportunity to disseminate jazz-specific content to a vast audience, including those who may not otherwise have access to jazz and would otherwise remain unaware of this musical tradition. This is made possible by the platform's functional logic, which is centered on virality. Users who have no or only a minimal connection to jazz are unlikely to be searching for jazz-related videos. Instead, they are exposed to this content as a result of algorithmic moderation. The user comments provide insight into how users perceive the musicians. They indicate that a significant proportion of young people who are active on TikTok are unfamiliar with jazz or exhibit a reserved attitude towards this music, which is often perceived as elitist. The interviewees unanimously agreed that their videos can facilitate low-threshold access to jazz for many users. They also asserted that this is crucial to ensure the future relevance and appeal of jazz to younger generations.

In conclusion, it can be observed that the image of jazz that is most visible on TikTok is contingent upon the affordances of the platform and the musicians' perceptions of the algorithmic functional logic and user preferences. A distinct algorithmic culture has emerged, which has resulted in the formation of platform-specific jazz-related performance conventions or vernaculars. Concurrently, the image of jazz represented on TikTok is contingent upon a multitude of external factors. This is particularly evident in how repertoire is selected, which is influenced by canonization processes that developed decades ago and are still at work today. The prioritization of normatively beautiful bodies is not a new phenomenon; rather, it has long been prevalent in popular music and media cultures, as well as other cultural domains. Moreover, the interview results indicate that established mass media, such as film and television, have not been abruptly rendered obsolete by platforms like TikTok. Rather, they influence the content on TikTok, for instance, by popularizing songs through series or movies, which then become trends on TikTok. TikTok is not a hermetically sealed digital space, but is strongly influenced by dominant cultural discourses and hegemonies. Nevertheless, the impact of platform algorithms and human engagement with these technical entities on cultural production and the associated cultural homogenization is a relatively recent phenomenon. The next chapter examines programmatic implications of these empirical findings for music research.

7. TikTok, Jazz, and Cultural Hierarchization

The pursuit of maximum media visibility, the dominance of relatively homogeneous groups of people and body standards, and a strong North American-centricity – some of the results of this study may initially appear unsurprising and even obvious when considered in the context of modern music and media cultures. Nevertheless, the representation of jazz on TikTok differs significantly from the conventional portrayal of this musical culture, which is dominated by the figure of the "jazz icon" (Whyton 2010). Jazz icons are a select group of male African American instrumentalists who are in many cases credited with having made a significant contribution to the further development of jazz over the last decades. They play a prominent role in the vast majority of accounts of jazz history. The legendary status of these musicians is justified by the core aesthetic criteria of musical virtuosity, complexity, and innovation. This understanding of jazz as a male-dominated African American form of art music suggests that it is not compatible with a broad public appeal or proximity to popular music cultures (DeVeaux 1991). It should be noted that this is only one of the many facets of jazz. However, it has long been prioritized in more traditional media contexts, including the book market, the magazine segment, and documentary films. It has also been prioritized in jazz education at conservatories and in jazz research, which has led to its naturalization as a highly representative variant of jazz.

In contrast, on TikTok, it is primarily young, *white*, and female singers who are successful with popular songs or jazz standards from the Great American Songbook repertoire. This indicates that in the TikTok context, facets of jazz are brought to the fore that deviate from the canonized and supposedly *official* image of this musical culture. Concurrently, the emphasis on particular demographic groups on TikTok serves to perpetuate cultural hegemony, a phenomenon that has long been pervasive in society, extending well beyond the context of jazz. The findings of this study demonstrate that the portrayal of jazz on TikTok is significantly influenced by the socio-technical interactions between musicians and the platform. These complex networks of human and non-human actors result in shifts in emphasis in terms of how much visibility is afforded to specific elements of jazz. However, these changes do not negate the continued prevalence of well-established cultural hegemonic structures.

In the course of this study, it became clear that on TikTok, only a small section of the diverse spectrum of what jazz can be and who can play it is singled out for increased media visibility. The dominance of comparatively old musical repertoires on the platform results in constant references to the past, even though the public perception of TikTok is primarily as a platform for young people where the principal focus is on current trends. Processes of memory production are thus documented on TikTok, insofar as music from the first half of the twentieth century is cultivated on the platform. Additionally, TikTok demonstrates the outcomes of cultural canon-building processes, in the course of which specific repertoires and groups of individuals become more prominent than others. It is evident that well-known hegemonies of social categories of difference and body norms play a pivotal role in this context. Similar processes have been occurring for centuries in a variety of cultural contexts, including music. However, in the context of TikTok, these cultural hierarchization processes are shaped by the introduction of algorithms and datafication under the conditions posed by platformization. This is a result of socio-technical interactions between human and non-human actors, which has led to the re-contextualization of these processes.

In the following, it is important to clarify what specifically is new about the processes of cultural hierarchization on TikTok presented in this study – and what is not. This is a crucial point: the extent to which such processes are transformed or possibly simply perpetuated on TikTok can only be measured in relation to long-established music-related logics of hierarchization or homogenization. Such a historical contextualization of current media phenomena is therefore – in addition to an empirical basis – a crucial prerequisite for understanding the actual culturally transformative potential of digital platforms (Hesmondhalgh 2022).

What distinguishes the jazz-related processes of cultural hierarchization on TikTok from comparable processes that have occurred outside of contemporary digital spaces in the past? What changes in focus occur when media visibility is no longer primarily controlled by traditional gatekeepers (such as mass media, including print, TV, and radio, as well as music industry players such as record labels), but is increasingly dependent on the interaction between human (primarily young musicians who are not necessarily integrated into professional music industry structures in the traditional sense) and non-human actors (the algorithmically controlled moderation logics of the platforms) in digital spaces?

7.1 TikTok and Digital Memory Production

Many of the songs from the Great American Songbook that have achieved considerable popularity on TikTok were composed and first recorded between the 1920s and the 1950s and have been a part of the jazz scene for decades. For example, they continue to serve as a foundation for improvisation among jazz musicians, particularly in jam sessions, and are subject to constant reinterpretation and re-recording. The *real books* that are commonly used in jazz contain lead sheets of a multitude of such songs, which are also referred to as *jazz standards*. It is evident that these real books reflect assessments and hierarchizations regarding which compositions are deemed to be *the standard* within the context of jazz and should, as such, be known to jazz musicians (Michaelsen 2013; Williams 2023). The act of repeatedly performing these songs serves to reinforce the creation of references to the past. This phenomenon is also evident on TikTok, where the selection of repertoire by popular jazz musicians maintains the visibility of comparatively old repertoires. In this sense, TikTok can be considered a medium through which certain musical repertoires are preserved in the public consciousness.

Memories are used to refer to knowledge about the past, i.e., to events that were either experienced personally or are understood as being culturally formative. Processes of visualizing the past can therefore relate to both personal and social history and can be enacted both individually and collectively. Fundamentally, memory processes entail an attempt to preserve the past in the present and to activate knowledge about the past (Lohmeier, Schwarzenegger, and Schreiber 2020, 49). On a supra-individual level, societies reaffirm their values and norms by remembering. That which is remembered is therefore primarily that which appears to have value for many people or which is defined as being worth remembering by actors with discursive power (Jost 2018, 84–85). With regard to the preservation of musical repertoires, supra-individual memory processes are involved. As with all memories, the memory of certain repertoires must be continuously refreshed if they are to remain meaningful for many people. Songs are thus not automatically embedded in the memory of a society, rather, specific bodies of knowledge must be actively accessed in order to be included in a cultural debate on memory. Such memories are transmitted from one generation to the next; they are modified whenever necessary and constantly renegotiated (Pentzold, Lohmeier, and Kaun 2023, 1–2). The popularization of older musical repertoires on a comparatively new platform such as TikTok leads to a renegotiation of existing memories. Transferring these songs to the TikTok context makes them accessible to a younger demographic who may not initially identify them as jazz standards from the first half of the twentieth century. Older repertoires are essentially preserved in this manner, transcending individual experiences and becoming an integral part of the cultural memory of Western societies (Menke and Grittmann 2023, 75–76). It is relatively simple to make these pieces accessible to supra-individual memory processes, given that they are already firmly embedded in the collective memory of society as jazz standards (Menke and Birkner 2023, 500). In contrast, an innumerable number of other memories are stored in archives, for example, and are not directly accessible to collective memories. Consequently, the past is inevitably recalled selectively and non-exhaustively (Pentzold, Lohmeier, and Kaun 2023, 2).

The media exposure of these songs and standards on TikTok has the potential to ensure that the status of these repertoires in the culture of memory will not diminish in the foreseeable future. Media in general are a fundamental prerequisite for societies' ability to record and communicate their memories. The role and influence of media in the formation of memory cultures is subject to historical processes of change. For example, the various and constantly evolving types of communication media have consistently shaped the dissemination of memories and the scope of what can be remembered (Burkey 2020, 181; Lohmeier, Schwarzenegger, and Schreiber 2020, 49). Consequently, digital media also give rise to distinctive forms of individual and collective remembrance that are inextricably linked to the media technology characteristics that are inherent to them (Pentzold 2021, 2). It can be argued that platforms like TikTok have the potential to influence processes of memory production, as they create certain media frameworks within which references to the past can be produced and communicated. However, it should be noted that the visualization of a music-related past, as documented in the case of jazz on TikTok, does not necessarily imply that the content creators explicitly intended to valorize a specific jazz-related past. Nevertheless, the interviewees Erny Nunez and Kellin Hanas in particular assert that they seek to preserve specific musical repertoires and make them more accessible to a younger demographic (see chapter 6.9). Ultimately, however, the selection of repertoire is also influenced by the prevailing popularization mechanisms that are characteristic of TikTok, as evidenced by a multitude of statements made during the course of the interviews.

It is therefore indisputable that particular forms of jazz-related memory production occur on TikTok. The extensive reach of TikTok and other digital platforms, which can also influence the social construction of memory, raises the question as to what extent platforms can either reinforce or potentially supersede the influence of traditional memory institutions. Upon initial observation, digital platforms appear to lack the fundamental archival premises that are characteristic of traditional archives. Archives are deliberately structured entities, and the individuals responsible for managing them have the authority to determine which documents or objects are retained or discarded. This enables them to exercise control over the potential for social recollection (Pentzold 2021, 5). Despite the absence of explicit disposal and storage practices on platforms, platform companies do exert efforts to control, curate, moderate, and monitor the content that circulates on their platforms. Digital objects, including photos, videos, and music files, are stored and made accessible to specific user groups, and the activities of users associated with these cultural objects are meticulously documented. In this sense, platforms can be considered archival infrastructures whose functionality is contingent upon automated processes of data collection, storage, and evaluation (Pentzold, Lohmeier, and Kaun 2023, 17). This is why specific platform mechanisms such as datafication and algorithmization (see chapter 3.2) inevitably intervene in processes of memory production in digital spaces.

It can be argued that platforms have the potential to supplement the cultural discourses of remembrance that are typically controlled by traditional memory institutions, such as archives and museums. Indeed, there is a possibility that they could even compete with these institutions. In the context of popular music cultures, alternative instances of memory work have already emerged and become established, particularly since the turn of the millennium. This primarily applies to archives run by private individuals (cf. e.g., Baker 2015; Baker 2017). The first jazz archives emerged in the early 1950s when private collections amassed by jazz fans were made available to the public (cf. Fitzgerald 2012). While state-funded museums and archives are often accused of approaching the subject of popular music culture in a relatively unreflective manner, sometimes reproducing common *heroic narratives* (Brandellero and Janssen 2014, 236), it is precisely the deconstruction of such narratives that is often the goal of private archivists. Since the early 2000s, for example, a number of feminist archives with a focus on popular music cultures have been established, which attempt to leverage the potential of independent online archiving (see Reitsamer 2018; on the topic of archival collections and gender, see Eichhorn 2013; Geraghty 2014, 53–71). These archives and their respective content areas constitute a counterbalance to the archiving practices of state-run memory institutions. This is particularly evident as the individuals employed in private archives endeavor to create more visibility for musicians and repertoires beyond conventional canons.

When compared to traditional memory institutions, such as museums and archives, platforms can facilitate greater accessibility to historical references for a broader audience. Unlike traditional institutions, which are often perceived as being inaccessible to younger generations, platforms like TikTok are embedded in the digital media experiences of young people. Their user-friendly interfaces enable low-threshold access to past-related content and actively engage younger audiences in memory culture discourses. Access to traditional memory institutions is typically associated with higher barriers, and many individuals are not well-versed in these institutions. While the influence of platforms may in principle result in memory discourses becoming more diverse, active participation in these discourses remains contingent upon specific resources and skills. Those who wish to actively participate must possess the requisite devices and media skills, which can be conceptualized as *mnemonic capital* (Burkey 2020, 182).

In the context of digital spaces, we are inevitably faced with the question of which actors and instances have the capacity to determine which references to the past are visible and to what extent, and subsequently find their way into digital memories (Pentzold 2021, 12). It is self-evident that platforms are not neutral actors. The datafication, algorithmization, and logics of selection and curation that underpin their functioning largely determine which objects and topics can become part of the negotiations of cultural memory within the context of the platform (Menke and Birkner 2023, 500–01). Thus, processes of selection and hierarchization of cultural objects and people, long reserved for human gatekeepers (such as archivists and museum curators), are increasingly being pervaded by algorithmic actors, themselves dependent on the logics of data aggregation and monetization of global platform companies (Collie and Wilson-Barnao 2020, 182). The economic and technological mechanisms that are fundamental to the functioning of platforms (see chapter 3.2) thus have a profound influence on the representation of cultures, the canonization of cultural objects and persons, and ultimately the production of memory in digital spaces. In addition, content creators' interpretations of the affordances of platforms and their corresponding imaginaries and strategies of action have a decisive influence on the production of the content that is disseminated in digital spaces.

Thus, technological advances in recent years and new forms of publicity in digital spaces have expanded the repertoire of media memory practices. This has enabled the establishment of new memory discourses, in which diverse memories can become visible without the intervention of traditional gatekeepers such as journalism or memory institutions such as museums and archives (Menke and Birkner 2023, 497). Academics, archivists and curators are therefore by no means in control of the interpretation of memory-related discourses. An exclusive focus on the memory work of these institutions and actors is therefore no longer sufficient if we are to understand the complex functioning of mediatized memories in contemporary media cultures (Lohmeier, Schwarzenegger, and Schreiber 2020, 59). It is therefore imperative to view platforms as a significant and evolving domain of memory production, one that engenders novel memory practices and challenges traditional forms of memory culture. The integration of TikTok into the media practices of a vast number of individuals, particularly young people, gives rise to the assumption that the forms of jazz-related memory production that take place on TikTok can also be effective beyond the platform. This particular mode of memory production is largely independent from entities like specialized archives, jazz journalism, and jazz research. In the future, music research will have to address such shifts in focus in order to ascertain which actors can influence the public image of music cultures or the communication of music-related memories, and to what extent.

7.2 TikTok and Musical Canon Building

As discussed in the previous section, one of the things that is negotiated in the context of memory production is who and what is considered culturally significant over a longer period of time and therefore also worth remembering. This phenomenon bears resemblance to the hierarchization processes commonly referred to as *cultural canon building*. This term encompasses the mechanisms through which influential cultural actors assess the value of cultural objects and individuals, determining their memorability and, subsequently, their visibility in the media.

The term *canon* has its roots in religious discourse and originally meant *standard*. It initially referred to those sections of the Bible that were deemed authoritative (Gabbard 1995, 3; Appen, Doehring, and Rösing 2008, 25). However, canon-building processes have long been discussed outside of theological contexts, including in music research. In this context, the criterion of authenticity is not the primary consideration. Rather, a canon in music-related contexts should be understood as a compilation of individual pieces, songs, albums, or musicians that are considered highly relevant within the context of their respective musical frame of reference. Individual cases are thus extracted from the heterogeneous totality of certain musical repertoires, such as the diverse forms of jazz, as being particularly worthy of being listened to, significant, valuable, and so forth. The selection process is always retrospective, and the confirmation of such a selection, particularly by influential figures such as music journalists, historians, and museum curators, enables the establishment of a canon within specific music cultures (Appen, Doehring, and Rösing 2008, 26). In view of the vast quantity of musical works that are currently available, canons provide a sense of security, stability, and orientation. For instance, ideas about significant musical compositions are transmitted from one generation to the next, resulting in a consensus among many individuals regarding these compositions and their sustained popularity over decades. While canons are not immutable, they do often possess a robust and enduring core over time, serving as a reference point for subsequent generations and their musical creations (Whyton 2013, 42–43). The results of such music-related canon-building processes are documented in a variety of sources, including journalistic best lists, recommendations for listening in reference books and specialist works, and academic presentations on the history of popular music.

Canon-building processes and the critical reflection thereof have been a core topic in jazz studies since the early 1990s (DeVeaux 1991; Gabbard 1995). Jazz-related canon-building processes, as are typically constructed and perpetuated by music journalism or jazz historiography, for example, usually attempt to establish jazz as an art music, which sometimes serves to legitimize this musical culture – as the American counterpart to European art music, so to speak (Gabbard 1995, 2). The supposed genius of individuals, primarily male instrumentalists, who more or less single-handedly brought about the great innovations in jazz, according to the typical narrative, is often emphasized. The musicians generally highlighted in this way form an extremely homogeneous group – female musicians, for example, are largely or even almost completely absent from almost all canonized accounts of jazz history (Whyton 2013, 43).

The exclusionary and homogenizing logics that generally underlie canon-building processes are therefore problematic. In music-related contexts, this refers not only to musical repertoires per se, but also to the authors or performers of these repertoires. Canon-building processes usually represent only the musical creations of certain groups of people, while others usually remain in the background or are even hidden due to social categories of difference such as gender, ethnicity, *race*, class, nationality, or geographical origin (Whyton 2013, 43). In an analysis of numerous rock and pop album lists, Ralf von Appen, André Doehring, and Helmut Rösing, for example, show that these lists are an extremely homogenizing form of musical canon building. The top positions on these lists are almost exclusively occupied by *white*, male musicians from the U.S. or Great Britain who play in a classic rock band line-up (vocals, guitar, bass, drums) and whose best-known albums were released in the second half of the 1960s (Appen, Doehring, and Rösing 2008, 33). The composition of such lists is typically the domain of music critics. It is noteworthy that the albums prioritized in these lists are not necessarily those that enjoy the most commercial success (Appen, Doehring, and Rösing 2008, 36). Consequently, the aesthetic attitudes of the discursively influential actors in the field of music journalism may not fully reflect the actual interests of listeners. The clear divergences that emerge from a direct comparison are likely attributable to the homogeneity of the group of individuals responsible for compiling such lists. It seems reasonable to posit that, at least according to von Appen, Doehring, and Rösing, these are *white*, middle-aged, male journalists with a comparatively high level of education. The music prioritized in the lists may be familiar to them for primarily biographical reasons, making it likely to offer potential for identification (Appen, Doehring, and Rösing 2008, 37).

Canon-building processes can frequently be observed to align with the cultural values of specific discourse leaders or cultural *elites*. This phenomenon is not exclusive to the domain of music journalism; it is also evident in the curricula of music-related academic institutions, which are largely dependent on the biographical influences and related cultural preferences of their academic teachers (Gabbard 1995, 3). The discrepancy between the music-related preferences that are actually widespread in society and the attitudes of these elites demonstrates that canon-building processes do not develop a universally valid normative force. Rather, they are only ever effective in certain areas of society. Nevertheless, canons such as those constructed in the field of music journalism can be disseminated, become socially sedimented, and exert influence over a significant number of individuals (Appen, Doehring, and Rösing 2008, 45–46).

Prior to the twentieth century, the canonization of music was predominantly associated with what is commonly referred to as art music. Consequently, while songs from popular music cultures may have achieved considerable recognition, they were not regarded as *serious* or as having the requisite quality to be considered *art* or *valuable* music worthy of preservation as part of the canon. However, throughout the twentieth century canon-building processes have expanded to encompass popular repertoires, as evidenced by the emergence of formats such as the aforementioned best lists (Appen, Doehring, and Rösing 2008, 30–31). Von Appen, Doehring, and Rösing identify three fundamental categories of canon building: the song canon, the musician canon, and the album canon. Within the context of the song canon, enduring songs, or *evergreens*, are preserved through various means, including radio broadcasts and the work of cover bands. In these instances, the focus is not on the creators or performers but on the songs themselves. In contrast, the canon of musicians is maintained through the medium of museum exhibitions, wherein a limited cohort of individual musicians and their oeuvre are repeatedly showcased and celebrated in a similar manner. The canon of albums is the most analogous to the canon-building trends observed in the domain of European art music and is associated with cultural capital. In music journalism, for instance, selected albums are designated *masterpieces of rock history*, which subsequently informs the construction of these lists of the most exemplary albums.

The prioritization of specific musical repertoires on TikTok, as evidenced in this study, aligns most closely with the logic of song-centered canon building. This is because, as revealed by the interview results, the creators of the Great American Songbook songs are just as obscure as the numerous albums by various musicians on which versions of these songs were released. The Great American Songbook songs that are so popular on TikTok are not necessarily linked to their composers or specific performers in the public perception, which is why a form of canon building that is not strictly historically oriented is evident in the context of these songs. At most, performers such as Frank Sinatra and Ella Fitzgerald, whom the interviewees associate with specific songs but who are not the composers, are mentioned several times in the interviews. The results of the interviews indicate that jazz standards such as "Fly Me to the Moon" are of great significance to a vast number of individuals, not only to the musicians who were interviewed but also to TikTok users. However, these standards do not occupy a prominent position within the jazz-related canon that has been constructed by those who are regarded as the elites. In any case, the repertoires preserved on TikTok are specific repertoires of jazz standards from the 1920s to 1950s. These standards were already popular prior to their distribution on the platform, and some of them were initially transferred to the TikTok context due to their use in more traditional media formats, such as films and television series. Consequently, TikTok serves to perpetuate and further popularize what was already pervasive in society in general.

The processes of cultural hierarchization are thus documented on TikTok and demonstrate parallels to earlier processes of music-related canon building. However, these processes are not congruent with those observed in the past. In the context of the platform, such processes occur under the influence of new media and new economic conditions. This means that TikTok does not produce top lists or comparable canon-building processes in which the same musicians and repertoires are prioritized again and again and with a high degree of continuity, which means that future canon-building processes can be prefigured to a certain extent. On TikTok, a video can be swiped directly if it is not to your liking; the brevity of the videos and the logic of the For You page encourage the consumption of many different videos within a relatively short time. In view of this, TikTok appears to be extremely ephemeral and is ostensibly unable to function as an instance of cultural canon building, since canon building is characterized precisely by the constant and long-term preservation of cultural hierarchies. Conversely, the very standards that have achieved popularity on TikTok in the context of jazz have for decades gained considerable recognition beyond the platform. Indeed, they were already canonized before their dissemination in digital spaces. This status as a *standard* can be further consolidated on TikTok, as evidenced by the repeated replaying of songs such as "Fly Me to the Moon" by innumerable users. Moreover, the videos remain accessible for an extended period, as they are stored on the platform and can be accessed at any time, provided they are not deleted by their creators.

It would appear that existing canon-building processes are being transferred to the context of TikTok, thereby increasing the likelihood of further social sedimentation. Musicians are acutely aware of the necessity of meeting the specific requirements of TikTok, and thus opt for specific musical repertoires in order to enhance their visibility on the platform. The interviewees were unanimous in their view that well-known and catchy songs, such as those from the Great American Songbook repertoire, are highly suitable for attracting the attention of users and encouraging them to interact. The affordances of the platform, the algorithmic imaginaries of the musicians, and their ideas about potential user reactions exert a considerable influence on the musical repertoires that receive the most attention on TikTok. This analysis does not focus on criteria that are typically applied in the context of jazz-related canon building beyond the platform, such as the artistic merit of the songs performed. However, processes of cultural hierarchization based on categories of social difference, such as origin and gender, which also substantially shape canon-building processes, are clearly reflected in the context of TikTok. Moreover, in contrast to its Chinese counterpart Douyin, TikTok is tailored to Western markets and evidently reflects a markedly Western-influenced, or more accurately, U.S.-centric perspective on jazz. This perspective, which is strongly contextualized within Western frameworks, has been a defining feature of jazz-related canon-building processes for decades, with U.S. repertoires and musicians typically accorded greater prominence.

7.3 Cultural Hegemonies on TikTok (and Beyond)

The findings of this study demonstrate that the homogenization tendencies observed on TikTok influence not only the musical repertoire performed, but also specific demographic groups. It is evident that body norms and beauty ideals play a pivotal role in this context. These social categories have historically exerted a considerable influence on the negotiation of media visibility, long before the advent of digital platforms. In such contexts, the human body should be understood as a medium for the construction and representation of social categories of difference, including but not limited to gender, *race*, and class (Degele 2006, 579). In this way, specific normative ideas of beauty, which are linked to social power, are inscribed in the body. Individuals who conform to specific, conventionalized beauty norms generally enjoy numerous privileges. The results of socio-psychological research demonstrate that individuals who are perceived as attractive can earn a higher income, find sexual and life partners more easily, and are perceived by others as being more likeable and competent (Degele 2006, 584). In this sense, the body can to a certain extent be used as capital to compensate for a lack of economic or cultural capital (Degele 2017, 116). This phenomenon is also evidenced by an analysis of the popularity peaks of jazz musicians on TikTok. The platform allows musicians to strategically utilize their bodies to gain popularity early on, circumventing the traditional gatekeepers of the music industry, such as record labels and managers. The results of the interviews show that, in addition to music-related aspects such as the choice of repertoire, the targeted staging of one's own body and thus also the perpetuation of normative ideas of beauty can significantly influence the likelihood of success or failure. It is evident that the category of *race* also plays a pivotal role in this context. As demonstrated in chapter 4.8, specific relations of inequality are reproduced on TikTok. BIPoC face significant challenges in generating long-term visibility on the platform. This power imbalance is also clearly reflected in the field of jazz popularity peaks on TikTok, where *white* musicians in particular are apparently afforded the opportunity to achieve substantial success on the platform. Specific body norms, particularly those pertaining to notions of attractiveness and the categories of *race* and gender, undoubtedly exert a profound influence on the negotiation of visibility on TikTok. Consequently, various categories of difference are inextricably intertwined on the platform, mutually reinforcing each other intersectionally under certain circumstances (for a detailed examination of intersectionality, see Collins and Bilge 2016; Winker and Degele, 2009; and Suzuki 2013, who offers a particularly insightful analysis of the phenomenon of intersectionality in the context of jazz).

The interviews conducted revealed that some musicians engage in targeted experimentation with their appearance, identifying specific patterns of success over time and subsequently presenting themselves in alignment with these patterns. This form of self-representation is thus tailored to the perceived demands of the TikTok platform, as imagined by the musicians themselves. Although these practices are clearly oriented towards the affordances of the platform, the underlying principle is not a novel concept. People have long engaged in social positioning through their appearance in a multitude of contexts, and they adapt their physical appearance to meet the requirements of specific occasions and contexts. Such beauty practices can be defined as work on one's own body and include any practices that aim to modulate one's appearance. These physical beauty practices are primarily associated with women and encompass makeup, hairstyles, and clothing, as well as measures related to hair removal and dieting (Goldmann and Herbst 2023, 939). The practice of social positioning through physical appearance can serve a variety of objectives, including establishing an identity and attracting attention (Degele 2006, 580). In the context of TikTok, attention is a valuable form of currency, which, as the findings of this study demonstrate, can be attained, among other means, through the presentation of particular physical characteristics. In particular, a platform like TikTok can contribute to the further perpetuation of specific body norms through its virality-centered and iterative functional logic. This is because successful TikTok creators can potentially serve as a benchmark for numerous users and aspiring creators, and their content has a greater chance of being noticed by many people on TikTok and subsequently being shared. Furthermore, the process of optimizing cultural objects for platforms can also be observed to entail the optimization of one's own body in accordance with established standards. Content creators who aim to achieve the highest degree of visibility are compelled to invest significant effort into enhancing their physical appearance (Goldmann and Herbst 2023, 940).

It is evident that the media has played a pivotal role in the representation of particular body norms even prior to the advent of digital platforms. This is because ideals of beauty and body norms have historically been negotiated and discursively produced within media spaces. However, digital platforms appear to merely replicate familiar homogenizations, as evidenced by current studies on popularity peaks on various platforms and in different content segments. A study conducted by Nicola Döring in 2023 revealed a striking gender imbalance in the top one hundred most subscribed YouTube channels. The author found that 75% of the corresponding YouTubers are male, while social media platforms such as TikTok and Instagram exhibited a more balanced representation of genders (Döring 2023, 964–66). In this context, YouTube reflects well-known gender stereotyping. For example, only five of the fifty most subscribed YouTube channels in Germany are run by individuals who identify as female. Additionally, the content of these channels also focuses on stereotypically feminine beauty topics. As Döring notes, no channel in the top two hundred and fifty German YouTube channels run by a woman addresses topics such as gaming, news, politics, or science, and only one channel is dedicated to sports (Döring 2023, 967). Well-known content creators who are active on platforms must establish connectivity within specific reference systems in order to be authenticatable for their followers. This often occurs within a heteronormative gender framework, which is why the channels of very high-reach female individuals are often about cosmetics, for example, while the channels of their male counterparts are about gaming (Schuegraf 2023, 959). Among the most popular content creators, there are only a few examples across platforms that deviate from these norms, such as individuals who create content related to the LGBTQ+ community or whose content is explicitly political in nature. The creators who generate the most attention appear to be those that reproduce gender-related differences and body norms that have become deeply entrenched in traditional mass media over decades and have ultimately been transferred to the platform context (Jost 2022, 416; Schuegraf 2023, 960). This gives the impression that, in the field of popularity peaks on digital platforms, everything is largely the same – despite the fact that, in principle, more people can participate in public discourse in digital spaces than ever before. This illustrates once again that, with regard to the most visible content and content creators, well-known cultural hegemonic structures are reproduced on digital platforms. Despite the diversity of voices that exists in principle, there is often a lack of democratic tendencies in digital spaces.

The portrayal of female jazz musicians on TikTok exhibits clear parallels to earlier media representations of women in jazz history. Some of the musicians interviewed for this study gained prominence on TikTok for their instrumental performances, particularly Kellin Hanas and Brooklyn Stafford. The remaining musicians occasionally play the guitar or piano to accompany their singing, but it can be assumed that they are still primarily perceived as singers. This example demonstrates that some established jazz-related patterns of canon building have remained consistent on TikTok. Sherrie Tucker has already performed a historical analysis of the "gender-coding of musical instruments" in jazz (Tucker 2002, 978–79). This refers to the historical fact that various instruments that have long been core features of jazz – such as drums, bass, and brass instruments – have always been typically associated with male musicians, while female musicians in jazz have essentially been accepted only as singers, if at all (see also Pellegrinelli 2008). The consequences of such gender-coding processes (and the corresponding omission of female musicians from the annals of jazz history) are still discernible today. This is apparent upon even cursory examination of the conventional overviews of jazz history, wherein at most a handful of female vocalists are juxtaposed with a multitude of male instrumentalists (see DeVeaux and Giddins 2015 as a typical example). These are representative of the gender-coded processes that are typical in the field of (music) historiography. The category of gender affects who can and may write history and who and what is (or can become) the subject of historical representations (Paletschek and Reusch 2013, 7). It is therefore unsurprising that jazz historiography, which has been heavily influenced by men from the outset, has always revolved around musicians who are perceived as male (Dunkel 2014).

The portrayal of female musicians on TikTok found in this study is analogous to the depiction of female jazz musicians in earlier media contexts. This phenomenon can be illustrated with the aid of a historical example: the popularity of all-girl bands in the U.S. between the second half of the 1920s and the immediate post-war period. As McGee (2009) and Tucker (2000) have shown, such bands enjoyed a considerable degree of popularity during this period. The ensembles were larger in size and comprised exclusively of female musicians. They typically performed in a conventional big band configuration that remains prevalent today. However, on occasion, they employed additional instruments, such as banjos and harps, for stylistic effect. One illustrative example is the band The Ingenues (McGee 2008). Many of these bands performed across the U.S. and were frequently reviewed in music periodicals such as *DownBeat* and *Metronome*. Despite their considerable success, they were occasionally subjected to sexist denigration in these media. During the Second World War, all-girl bands were on the one hand supposed to provide entertainment and variety, and on the other hand were formed to replace the numerous male musicians who were absent due to the war. Following the conclusion of the war and the return of the men, female musicians were – in some instances deliberately and systematically – excluded from the music industry with the objective of reinstating the roles of the (mostly) male musicians. Moreover, music journalists wrote disparagingly about female musicians, deeming them unworthy of artistic recognition and remembrance. Consequently, all-girl bands were excluded from the canon of conventional jazz history narratives (McGee 2009, 245–57).

Nevertheless, the work of these bands is relatively well documented in numerous short films. In these films, the female musicians are occasionally depicted as serious musicians, essentially as instrumentalists. In some instances, however, the women appearing in the films are not, in fact, musicians, a fact that often becomes evident from the awkward way in which they handle the instruments. In such cases, the music was played back from a tape. In instances where the emphasis is not on technical proficiency but rather on other aspects, such as in these short films, female musicians are often depicted in a stereotypical and sexualized manner. The staging is at times evidently derived from the then-popular pin-up girl aesthetic (McGee 2009, 134–67). The parallels in attire and cosmetics between the women featured in these films and the most prevalent female jazz musicians on TikTok are, at the very least, noteworthy. This example illustrates that the stereotypical and sexualized performances of female jazz musicians, as documented on TikTok, can be traced back approximately one hundred years in jazz history. Consequently, the representation of the most popular female jazz musicians on TikTok is part of a long-standing tradition that is being renegotiated and, to a certain extent, perpetuated on the platform under changing media conditions. In the context of new media, certain influencer logics exert a discernible influence on the representation of jazz. These logics posit that female-presenting individuals can achieve a high reach, particularly within the beauty and lifestyle segment (see above), and they portray themselves accordingly.

A further parallel is that the majority of all-girl bands whose work is documented in short films were comprised entirely of *white* musicians (with a few exceptions, for example The International Sweethearts of Rhythm). Despite the fact that jazz has consistently been shaped by a combination of African and European American musical traditions, it has frequently served as a vehicle for expression among Black musicians over the course of several decades. Nevertheless, since the early days of jazz documented on phonograms in the 1910s, it has been observed that *white* musicians, with the support of the music industry, have on occasion achieved considerable success. This phenomenon is not exclusive to jazz; it can be observed repeatedly in the history of popular music cultures. A case in point is the singer Elvis Presley, who was known as the *King of Rock and Roll*. Throughout his life, Presley was accused of appropriating Black R&B and achieving significant success with it, while the creators of the songs, the musicians who inspired him, and his songwriters remained largely in the background. In the context of jazz, similar accusations have already been made regarding the band that is said to have produced the first-ever jazz recording in 1917 – at least according to the usual account in most jazz histories. This group, the Original Dixieland Jazz Band, consisted exclusively of *white* musicians and came from New Orleans, the supposed birthplace of jazz, where there was indeed a tradition of *white* jazz bands. It is notable that the majority of jazz ensembles from this era that have left an enduring legacy in the history of jazz were ensembles comprising solely or predominantly Black musicians. The success of the Original Dixieland Jazz Band's recording, which was significantly influenced by the record company Victor, whose employees first came across the band in the New York nightlife scene and are said to have recognized its commercial potential (DeVeaux and Giddins 2015, 68–69), is still regarded as a pivotal event, both in terms of its cultural impact and its economic consequences. At the time, the music industry was shaped by a pervasive racial bias, which often resulted in the exclusion of Black musicians from recording opportunities and the marginalization of their work in radio programming. Consequently, jazz was often perceived as a form of popular music performed by *white* dance bands (Garofalo and Waksman 2013, 28).

In the subsequent period, the *white* bandleader Paul Whiteman, also known as the *King of Jazz*, achieved considerable commercial success with his *Symphonic Jazz*, which was heavily influenced by European orchestral music. This success, which endured until the 1930s, also led to the production of the feature-length film *King of Jazz* in 1930. With regard to the accusation that *white* musicians appropriated jazz, one scene towards the end of the film is particularly noteworthy: it features a text panel referring to the U.S. as "The Melting Pot of Music [...] wherein the Melodies of all Nations are fused into one great New Rhythm: Jazz!" followed by a compilation of performances by various dance and music groups. The groups are attired in matching costumes and perform different European and Euro-American music traditions. According to the film, these traditions all contributed to the development of jazz. Notably, the film makes no reference to the significant influence of African American music, instead presenting jazz as a purely *white* musical form. This example illustrates how, by the early 1900s, media-constructed caricatures of jazz history with regard to the category of *race* were already circulating. Further examples of *white* jazz musicians who, having achieved significant commercial success, were accused of exploiting African American music culture can be found throughout the history of jazz. These include well-known swing big band leaders such as the *King of Swing* Benny Goodman in the 1930s and 1940s (Tackley 2012), the pianist Dave Brubeck from the 1950s onwards (Klotz 2023a), and the smooth jazz saxophonist Kenny G, who emerged in the 1980s (Wright 2023; Klotz 2023b). We can also observe a similar phenomenon with singers such as Norah Jones and Diana Krall from the late 1990s onwards (Arndt 2006). These examples demonstrate how economically driven and medially disseminated inequalities between *white* and Black musicians have been a persistent feature of jazz history for over a century. These hegemonies are not a recent phenomenon but are currently being reproduced under the influence of contemporary media, including in the context of digital platforms.

Indeed, the phenomenon of *white* musicians attaining greater success than the original creators of African American cultural forms of expression has been a persistent and recurring theme in the history of popular music cultures for approximately two centuries. In their analysis, Reebee Garafalo and Steven Waksman identify the origins of these developments in the context of minstrel shows, which became a popular form of entertainment in the United States during the first half of the nineteenth century. These shows typically featured *white* performers, who racially parodied Black people on stage, often with the use of blackface:

> Minstrelsy established a vexing and recurring pattern of uneven musical exchange in which white interpretations and appropriations of African American culture would receive disproportionate credit in defining mainstream

> popular culture, while black performers would struggle for visibility even when black music or culture was being portrayed. (Garofalo and Waksman 2013, 17)

While racist parody ceased to serve as the foundation for such appropriations in subsequent decades, the fundamental tenets underlying these processes remained remarkably resilient.

The Great American Songbook repertoire, so popular on TikTok, is also a product of specific processes of appropriation and reinterpretation of African American musical forms of expression by *white* musicians. Well-known American songwriters such as George Gershwin, Irving Berlin and Cole Porter are generally considered to be representative of and influential for the repertoire of the Great American Songbook. As early as the early twentieth century, they wrote numerous songs for Broadway musicals and Hollywood films which are still well known today in connection with Tin Pan Alley in New York. It has been demonstrated that these *white* songwriters often engaged profoundly with popular African American music, as evidenced by George Gershwin, for example (see Noonan 2012, 148), who took lessons from ragtime pianist Luckey Roberts (Peress 2004, 67–68), among others. In particular, ragtime and early jazz constituted a significant source of inspiration for songwriters, as can be seen in the compositions of Irving Berlin, who himself produced several ragtime songs. The aforementioned songwriters achieved a breakthrough in the U.S. mainstream segment as a result of the influence of African American music culture, which was perceived by many contemporaries as alien (Garofalo and Waksman 2013, 25; Knauer 1990, 65). Notwithstanding the pervasive influence of African American music culture, the professional songwriting circles of the time remained a predominantly *white* space, largely due to the pernicious racial structures that permeated the music industry (Garofalo and Waksman 2013, 29).

The observation that the most prevalent interpretations of this musical repertoire on TikTok are predominantly performed by *white* individuals has the potential to engender a perception of jazz as a *white* space on the platform, not only in terms of visual representation but also with respect to the auditory dimension. The association of certain sounds with racial stereotyping is a phenomenon that has been extensively documented. For instance, specific phonetic and dialectal nuances have been identified as auditory markers of Blackness (Stoever 2016, 8). Over time, norms of timbre have emerged that result in voices being perceived as either clearly Black or *white* by many people

(Eidsheim 2019, 196). In a musical context, tonal nuances such as blue notes, timbre qualities (e.g., roughness), specific rhythmic patterns, and groove characteristics, which are commonly associated with genres influenced by African Americans (e.g., blues, funk, rap, jazz), can serve as markers. The most popular jazz performances on TikTok rarely feature these characteristics. This can result in the most prevalent jazz genres on the platform, which are significantly influenced by *white* artists, being perceived by TikTok users as the norm. This is where Jennifer Lynn Stoever's concept of "whiteness in an auditory sense" becomes evident: namely, the pervasive and unquestioning normalization or naturalization of *whiteness* at a sonic level (Stoever 2016, 12).

The research on racist and sexist biases in algorithmically moderated spaces has repeatedly shown that specific social inequalities are perpetuated in contemporary media cultures (see chapter 4.8). In this context, Ruha Benjamin coined the term *Jim Code* a few years ago, drawing a parallel between the minstrel show character Jim Crow and the discriminatory practices observed in digital spaces. Benjamin underscores the pervasive misconception that technologies are inherently neutral entities. This misperception contributes to the failure to recognize the role of algorithmic systems in perpetuating racist structures: "'[T]he new Jim Code': *the employment of new technologies that reflect and reproduce existing inequities but that are promoted and perceived as more objective or progressive that the discriminatory systems of a previous era*" (Benjamin 2019, 5–6; italics in original). It is important to recognize that racism can be viewed not only as a consequence of technological advancement, but also as a crucial element influencing the evolution of technology itself (Benjamin 2019, 40). On the one hand, self-learning algorithms are shaped by the worldviews of the individuals responsible for their development. On the other hand, the behavioral patterns of users, upon which the algorithms base their learning, also exert a significant influence (Benjamin 2019, 50). In this regard, sensationalist headlines, particularly prevalent in journalistic discourse, in the style of "Is the TikTok algorithm racist?" represent a manifestation of media determinist perspectives and are inherently limited in their scope. They imply that algorithms, in and of themselves, could be inherently racist (Benjamin 2019, 44).

As the examples of relations of inequality in the course of the history of popular music cultures presented in this chapter demonstrate, the events on TikTok analyzed in the context of this study and shaped by various biases have a number of historical antecedents. In this sense, history is, as it were, maintained in the algorithmically moderated spaces on the platform. However, as the research results presented here clearly demonstrate, this is not contingent on the platform's algorithms alone. Ruha Benjamin underscores the influence of socio-technical constellations in digital spaces with regard to the perpetuation of cultural hegemony, but particularly in relation to the actions and worldviews of human actors who shape the evolution of algorithms. As demonstrated in this study, the content creators' pursuit of visibility and their strategies also play a pivotal role in shaping the representation of music cultures on TikTok. The content creators are guided by two key factors in determining their strategies of action: their own algorithmic imaginaries and their experience and assumptions regarding how TikTok users respond to their videos. The TikTok universe is characterized by complex socio-technical relationships between the platform, its algorithmic system, musicians, and users. These relationships shape the image of jazz that emerges on TikTok, with all the heterogeneous actors involved, human and non-human, playing a role in this process.

7.4 Socio-Technical Canon Building

TikTok has the potential to structure actions and affords certain music-related content and stagings. Concurrently, cultural hegemony exerts an influence over the portrayal of jazz on TikTok, which has been a constitutive factor in the negotiations of media visibility in a comparable form for many decades. Moreover, jazz musicians who are active and successful on TikTok play an active role in the negotiations surrounding the visibility of individuals and content within the jazz context on TikTok through their interpretations of the functional logic of TikTok. Consequently, there are divergences from the predominantly academically constituted jazz canon that has been perpetuated for decades. This is evidenced by the absence of "jazz icons" (Whyton 2010) from the domain of popularity peaks in jazz on TikTok. This is primarily due to the fact that, according to the interviewees, these icons do not align with the platform's logic of popularization. Additionally, TikTok tends to prioritize specific body norms that young female musicians are more likely to conform to.

It is evident that on TikTok, specific aspects of jazz become more prominent than others due to the interplay between human and non-human actors. These processes are not solely determined by technology or exclusively shaped by society. Algorithmic sorting logic and platform-specific affordances influence the processes of cultural hierarchization, yet human actions and established cultural hierarchies continue to play a significant role. The consequence is a selective shift in focus with regard to the media visibility of specific facets of jazz. However, the groups of people and the musical repertoires that are popular on TikTok are not isolated from the tendencies towards canon building and homogenization that also exist beyond the platform. Although the young female musicians conform to the prevailing standards of beauty and can leverage their physical appearance as a form of capital in the digital landscape (Gugutzer 2022, 97–105), where visibility is paramount, the celebrated Great American Songbook songs and jazz standards occupy a distinguished position within the canon of musical repertoire (Michaelsen 2013). However, this canonization does not align with the conventional norms established by traditional gatekeeping institutions, such as jazz journalism and jazz research.

The interactions between platform mechanisms, user reactions, and musicians striving for success on the platform have led to the emergence of *socio-technical canon-building* processes on TikTok. These processes are not independent of established canon-building processes but nevertheless develop their own logic. TikTok is thus not a hermetically sealed digital space that produces cultural hegemonies independently of or in isolation from external influences. However, well-known hegemonies can be perpetuated and potentially intensified on TikTok due to the platform's homogenizing logic.

Recent developments demonstrate that the influence of individual platforms can indeed wane over time. For example, Facebook and X (formerly Twitter) are likely to be far less relevant to young people than Instagram and TikTok. Additionally, numerous former users have boycotted X for political reasons, primarily in the wake of the acquisition of the platform by Elon Musk. It seems inevitable that TikTok will lose its relevance in the near future. However, the socio-technically induced canon-building processes that are currently taking place on TikTok are not exclusive to this platform. Rather, they are a cross-platform phenomenon that can also be observed in modified form on other platforms such as YouTube and Instagram. Since the cultural hierarchies that are specific to TikTok are in many ways based on well-known canons, they have the potential to have an impact beyond the platform in the long term.

In light of these considerations, it seems plausible that the particular representation of jazz on TikTok may exert an influence not only on the platform itself but also, at the very least, on the social perception of this musical culture. The social establishment of new media has the potential to exert a profound influence on processes of social communication, thereby also affecting the construction of reality that is produced under the influence of specific media (Menke and Grittmann 2023, 72). Given that processes of cultural canon building, which have long been constitutive for various music cultures, occur in modified form on a platform as popular as TikTok, it is possible that they may also have a lasting impact on the perception of music cultures beyond the platform. The TikTok platform may not necessarily maintain its popularity for several decades or even to the late 2020s. Such a situation could change rapidly. The crucial factor is that what has already been canonized in society is made even more visible on TikTok, and that the platform actually plays a role in the everyday structures of countless individuals, especially young people. “In terms of musicality and artistry, [TikTok] really streamlines what people do,” as the interviewee Sam Ambers puts it – and this streamlining will probably have an impact on the reception practices of TikTok users to a certain extent. The increasing exposure of listeners to established repertoires, which in some cases pre-dates the advent of TikTok, may reinforce the canonization of these works. The considerable reach that popular jazz musicians can achieve on the platform is a key factor in this phenomenon. After all, TikTok is not a music magazine that is hardly read, nor a physical archive that almost no one knows about, let alone visits – it is one of the world's most popular digital platforms, and its use is firmly embedded in the everyday lives of countless people. That is why the cultural hierarchizations that are documented on TikTok can spread much more widely than those that originate from more traditional gatekeepers.

Nevertheless, the platform's vast user base does not ensure that the image of jazz emerging in the context of TikTok popularity will be perceived sustainably within the jazz community. It is plausible to assume that people working in the professional jazz sector, such as professional musicians and university educators, may not be exposed to the videos of the musicians interviewed for this study or comparable content. Disparate facets of jazz may prevail within the echo chambers of individuals inside and outside digital spaces. This would not be unexpected, given that canon-building processes are unable to achieve a universal impact (see above). Consequently, it is not reasonable to assume that under the influence of TikTok, all canon-building tendencies, some of which have existed and been maintained for decades, will suddenly become obsolete. Nevertheless, the jazz standards that are so popular on TikTok, such as “Fly Me to the Moon,” appear to be facets of jazz that are actually appreciated by many people, particularly outside of highly specialized academic circles. In this respect, the homogeneous jazz performances on TikTok have the potential to have a lasting effect on the broad social perception of jazz. Despite the superficial volatility that characterizes its current status, TikTok represents a significant potential for musical canon building in the twenty-first century. In this context, established patterns of cultural hierarchization are undergoing a process of recontextualization, shaped by evolving media conditions and the influence of diverse actors, including human creators and users, as well as non-human entities such as algorithms.

7.5 Jazz Research and Platformization

At this point, I would like to return to the statement by Haftor Medbøe and José Dias quoted in the second chapter of this study: "Jazz has been slow to embrace the power of social media and seems to consistently arrive late at the table be it in the examples of MySpace, Facebook and Twitter" (Medbøe and Dias 2014, n.p.). It is evident that this assessment is no longer entirely accurate at the present time. However, if it we modify it slightly and apply it to jazz research, it is still valid: "Jazz *research* has been late at the table." As previously discussed in chapter 2, there is still little research on jazz in the context of online media. Even in a relatively new and comprehensive international publication such as the *Routledge Companion to Jazz Studies* (Gebhardt, Rustin-Paschal, and Whyton 2019), there is a paucity of material exploring the relationship between jazz and digital media, with the exception of the contribution by Barber (2019) in this compendium. However, in the past three decades, jazz research, particularly within the context of New Jazz Studies, has identified a number of core topics that are closely related to the key topics developed in this study. This applies in particular to publications on the topics of gender (e.g., Knauer 2016; Rustin and Tucker 2008; Reddan, Herzig, and Kahr 2023), canon building (e.g., Knauer 2018a; Whyton 2010; Whyton 2013) and mediality (e.g., Heile, Elsdon, and Doctor 2016; McGee 2009). Within this context, the topic of jazz on TikTok, or on digital platforms in general, appears to be quite compatible with jazz research. Jazz is currently integrated into contemporary media cultures in a variety of ways, and the major topics of recent international jazz research, supplemented by approaches from platform studies, provide a solid foundation for further research into these connections.

Furthermore, TikTok provides insights into the production and reception of jazz-related content that reaches a vast audience. The expressions of jazz that are popular on TikTok attract considerable attention from a vast number of individuals, as evidenced by the TikTok videos' frequently sizeable number of views. However, these forms of jazz are relatively distant from the established characteristics of jazz that are typically recognized as such or elevated to the status of *good* jazz by gatekeeping authorities within the realms of classical academia. The analysis of jazz on TikTok offers insights into the jazz-related production and reception practices of individuals who do not necessarily have an academic background in jazz and who are not primarily interested in specific niche segments within jazz that are commonly considered particularly artful and therefore culturally valuable. Historically, there has not been much interest among jazz researchers in musical repertoires that are actually popular. Smooth jazz, for example, represents a variant of jazz that enjoyed considerable success in the 1980s and 1990s. It has had a significant and enduring impact on the prevailing conceptions of jazz among a vast number of individuals. It is conceivable that this is the fundamental or sole facet of jazz that the majority of individuals have ever been exposed to, or at the very least, the sole facet that they have deliberately engaged with (Washburne 2004, 134). Hybrid genres such as smooth jazz, situated at the nexus of jazz and popular music, have historically encountered significant challenges in both jazz and popular music studies, with studies of these genres having hitherto been largely overlooked (Ake, Garrett, and Goldmark 2012; Burkhart 2024). In this regard, the prevalence and popularity of jazz on TikTok aligns with the tradition of *light* jazz repertoires, which have been largely disregarded by jazz research thus far. However, contemporary jazz researchers must address such phenomena if they are to gain an understanding of the social relevance of jazz, its multifaceted nature, and its dissemination via diverse media channels. This can only be successful if the canonization of subjects typically associated with jazz is avoided and research subjects are chosen in such a way as to avoid the perpetuation of such canon-building processes.

The results of the present study demonstrate the emergence of jazz-related hierarchizations on TikTok, which appear to be largely independent of the influence of established gatekeepers in the field of jazz studies. At first glance, the platform appears to be operating with a high degree of autonomy. However, this observation only captures half of the story, as the analysis in this chapter has revealed that even within the context of a digital platform, the process of cultural hierarchization is not a radical departure from established patterns. What is certain, however, is that the actors and, to a certain extent, the mechanisms of cultural canon building have changed in the platform age. Reflecting on and understanding these processes and the associated cultural power relations must be the task of contemporary and critical research (not only) on jazz.

The present study, regardless of its specific focus on jazz, demonstrates that the conditions for particular representations of music cultures on TikTok can only be identified through detailed qualitative research. Consequently, there are inevitably some open questions and gaps in the research, given that qualitative research designs entail the analysis of selected individual cases and do not aim to produce results that can be generalized statistically. Qualitative and hypothesis-generating research is, however, a prerequisite for countering speculative attempts to explain the influence of digital platforms on music cultures with an empirical foundation and for fostering a better understanding of the *actual* cultural power of platforms.

Despite the recent increase in publications on the topic of music and platformization, numerous core questions remain unanswered. This is particularly evident with regard to the specific ways in which musicians navigate the novel technological and media conditions offered by digital platforms. It is evident that profound shifts in media have consistently shaped cultural production, including in the domain of music. Consequently, it is reasonable to hypothesize that distinctive platform effects (Morris 2020) may emerge over time (see chapter 3.6). However, the precise manifestation of these platform effects is not dictated by digital platforms in a deterministic manner. Instead, it is contingent upon the creative endeavors of music creators. Digital platforms are unlikely to have a directly comparable impact on musical production processes across genres. While it seems perfectly plausible that songs in the mainstream pop segment are – at least to a certain extent – formally adapted to conform to the popularization logic of platforms such as Spotify and TikTok, this probably applies to a much lesser extent – if at all – to black metal or modern jazz productions, for example. It is therefore highly improbable that the fundamental essence of music itself will undergo a transformation as a consequence of the influence of digital platforms. There is a pressing need for qualitative empirical studies that focus on specific music-related production cultures in the context of digital platforms, with a detailed examination of how musicians respond to the technological and media upheavals that have occurred in recent years. Furthermore, the findings of the present study indicate that popularity on TikTok is not solely contingent on musical elements. Instead, it is also shaped by social categories such as *race*, gender, and specific body norms. Explanatory approaches that prioritize musical aspects and fail to consider the multifaceted cultural contexts of the music under study fall short in this regard.

The present study is unable to provide definitive answers to the question of what, if any, concrete influence the representation of jazz on TikTok has or can have on the perception of this music culture beyond the confines of the platform. In the interviews, some of the musicians indicated that their activities on TikTok are not representative of their musical work outside the platform. Moreover, it is currently unclear to what extent other jazz musicians may be influenced by jazz performances on TikTok. Additionally, while the behavior of TikTok users demonstrably plays a role in the platform-related imaginaries of musicians, the modes of reception of fans have inevitably been excluded from this study due to the chosen focus on the sphere of cultural production. However, to gain a deeper understanding of the intricate socio-technical networks within the platform environment, it is essential to conduct qualitative empirical studies on reception practices, as these can also impact the musicians' production strategies. For instance, it would be valuable to investigate whether the musicians' strategies of action, as empirically reconstructed in this study, align with users' expectations. Furthermore, it would be beneficial to ascertain the degree to which users employ the platform algorithms to influence the content they are shown. By establishing a foundation for reception studies based on platform-related production research, we can gain valuable insights into the role played by digital platforms in shaping contemporary music and media cultures.

The findings of this study indicate that a certain degree of music-historical contextualization is indispensable for the interpretation of the platform-specific representation of jazz on TikTok. This is particularly evident with regard to the cultural hegemonies prevalent on TikTok and, correspondingly, with regard to the jazz repertoires popular on this platform. The logic of cultural hierarchization can be seen to remain remarkably stable over extended periods of time, even when new influential gatekeepers, such as digital platforms, emerge. However, there has been a tendency over recent years to portray music-related phenomena in the context of digital platforms as entirely novel, particularly in journalistic discourse as well as in academic publications that reference these discourses. In doing so, an attitude informed by a general critique of contemporary culture is sometimes adopted, in that platform-related technological innovations and their supposed influence on practices of music production and reception are subjected to harsh criticism (on cultural criticism in general cf. Bollenbeck 2005; Hecken 2016). This phenomenon can be observed, for example, with regard to the chill playlists that are very popular on Spotify and the supposedly light tracks that are compiled on these playlists and are suitable for background listening. In this context, Spotify is sometimes criticized for encouraging the development of passive music reception practices and prioritizing what is often referred to as *functional music* (for a journalistic perspective on this, see Pelly 2017 and Pelly 2018, and for a critique of this argument and its adoption in academic publications, Hesmondhalgh 2022). Regardless of the questions of why – firstly – certain music-related reception practices are deemed better than others and whether – secondly – Spotify playlists are actually used in the way the authors assume, such critical statements are simply ahistorical. Comparable accusations have been made in recent decades with regard to easy listening, muzak, and smooth jazz, i.e., so-called *light* or *functional* music. The history of these musical genres has been the subject of academic inquiry by numerous scholars (see, for instance, Keightley 2008; Sterne 1997; Jones and Schumacher 1992; West 2008; for an overview, see Hesmondhalgh 2022, 9–11). Consequently, the corresponding criticism of their purported purely *functional* character is also well documented. Thus, while TikTok draws on decades-old Great American Songbook songs, the *mood* and *chill* productions and playlists on Spotify are significantly influenced by musical developments that occurred during the latter half of the twentieth century. The history of what is commonly referred to as functional music, of course, extends much further back in time. The aesthetic debates on autonomous and functional music first peaked during the nineteenth century (de la Motte-Haber 2017). The production of supposedly functional music and the criticism leveled at corresponding music forms are thus not a new phenomenon. And the much-cited platform effects (Morris 2020) must always be interpreted in light of earlier economic, media and technological contexts that influenced music production and recording processes, the results of which have been described by Mark Katz as phonograph effects (Katz 2010).

The insights that can be gained from an ahistorical perspective on platform-specific phenomena are very limited in scope. To reach a comprehensive understanding of the phenomena emerging in the context of platforms, it is essential to consider the historical trajectory of such phenomena. This allows for the identification of instances where a phenomenon has been canonized and perpetuated, as well as instances where there have been shifts in emphasis. The development of platform-related music cultures is shaped by two key factors: the individuals who actively engage with these platforms and the algorithms that drive their functionality. The individuals are influenced by a complex interplay of cultural and historical patterns, while the algorithms rely on data accumulated over centuries.

While platforms themselves indicate their potential for exploration due to specific research affordances (Burgess 2021; see chapter 5.1), certain researchers only consider certain platforms as potential objects of study. Sometimes the reasons for this are mundane; language, for example, plays a crucial role. Given my proficiency in only two languages, namely German and English, at an academic level, I was forced to constrain my research to Western platforms. A comparison with the representation of jazz on TikTok's sister platform Douyin would have been a logical choice but would only be feasible in a research team with the appropriate language skills. Consequently, my study reproduces the specific imbalances of Western or Anglophone platform research, which for the most part focuses on the Western market. Notwithstanding the critical reflection and empirical justifiability of my selection of cases for analysis, it is inevitable that I reproduce specific relations of visibility in the platform context by focusing my research mainly on *white*, predominantly North American content creators. This study will need to be complemented by research that does not concentrate on the popularity peaks in the jazz genre on TikTok. Consequently, it will be necessary to consider the perspectives of those groups of people who were excluded from my work due to the specific research question and the corresponding research design. The results presented here naturally raise new questions that can be examined in the context of further detailed studies. In the long term, the goal must be to research as many music-related spheres as possible in the context of platforms in order to create comparability and to do justice to the heterogeneity of music cultures in digital spaces.

8. Coda

As I conclude this study in the summer of 2024, approximately two and a half years have elapsed since I began my academic investigation of jazz on TikTok. Not a long time in academia, one might think. In fact, for various reasons, large scientific projects sometimes take years to complete. The ideal academic paper is characterized by scholarly rigor, reflection, transparency, and diligent handling of sources. Additionally, academic work is a lengthy process due to the necessity of developing and modifying questions and research designs throughout the process and – when necessary – discarding and redesigning them. Typically, a significant amount of secondary literature must be sourced and read. In qualitative empirical research projects, it is for example necessary to identify and contact potential interview partners, who must then be convinced of the merits of participating in the research project. The data collected during the course of conducting scientific interviews must be evaluated and interpreted using rigorous procedures that meet scientific quality criteria. Only then can the findings be documented, which requires a significant investment of time, particularly in the case of academic monographs, which often span several hundred pages. These steps must be undertaken concurrently with the inherent responsibilities of academic work, which can be equally time-consuming. Such responsibilities include academic teaching and self-administration, committee work, supervising theses, holding lectures, participating in conferences, acting as a reviewer, and engaging in other research projects. The next step, the publication of one's research results may also be a protracted process, as publications are frequently subjected to a review process by individuals who are also active within the scientific community and may be constrained by competing demands on their time. In the case of book publications, securing funding is often a prerequisite for publishing with a specialist publisher, given the significant costs involved. Following the submission of a manuscript to a publisher, the publication of the book may take several months or even years. In addition to these factors, the precarious working conditions in the academic world must be considered. Researchers – at least in the German-speaking countries where I work – typically only receive fixed-term employment contracts, which means that they have to constantly apply for jobs and are often forced to relocate or commute long distances. These circumstances are highly unfavorable for focused work on complex scientific topics. There is often a lack of tranquility and composure. And then there's this thing called private life...

It is possible that those outside the scientific community may be largely unaware of these processes. However, it is unlikely that any individual within the scientific community will be surprised by the above description. These are the rules of academia, you might object, and perhaps dismiss the above paragraph as self-indulgent griping. With a topic such as the present study, however, it is by no means insignificant that the production and publication of research results can sometimes take years. This is due to the fact that the topic of digital platforms, and in particular TikTok, is one that is undergoing rapid and significant changes. While the fundamental trends in the representation of jazz on TikTok are likely to endure in the longer term, the results of my corpus analysis, which was conducted at the beginning of 2022, no longer reflect the current state of affairs. Some of the musicians interviewed for this study still rank among the most popular jazz-related content creators on TikTok. However, the results of an analysis conducted at this juncture would likely differ, at least in part, from those presented in this study. For the reasons previously stated, the preparation and publication of a follow-up study would require a significant investment of time and necessitate the immediate initiation of yet another follow-up study. This cyclical process would continue indefinitely, rendering it impractical for scientists to keep pace with the evolving subject matter of their research. The rapidity of change in the field of TikTok, coupled with the inherent limitations of the scientific process, makes it challenging to stay abreast of new developments.

All of the musicians interviewed for this study still maintain an active presence on TikTok, although their activities have undergone some modification. In certain instances, the frequency of their posts has diminished in comparison to the levels observed at the time of the interviews. Moreover, the average number of views their videos receive is no longer on par with the figures they consistently attained two years ago. Some of the musicians now prioritize their presence on Instagram, while the content of their videos has also undergone a transformation. Sam Ambers, for instance, has shifted his focus to comedy videos, which he produces in collaboration with other content creators. His latest videos rarely make any reference to jazz.

The reality that scientific discourse is unable to keep up with a rapidly evolving phenomenon such as TikTok, due to the presence of established and (at least in most cases) necessary processes in academia, represents a challenge that is likely to persist. This situation can, at times, lead to feelings of frustration, as evidenced in the context of academic teaching. I have previously taught courses on music streaming, with a particular focus on Spotify, and on music cultures on TikTok. In these courses, I was able to provide students with studies on Spotify and TikTok, the results of which were however based on data that was, in all cases, several years old. This was due to the circumstances previously described. A review of the existing literature revealed a dearth of information on the contemporary contexts of music production and reception.

All of this is sometimes very unpleasant. But does the comparatively long time that has elapsed between the start of the research work and the final written form diminish the fundamental significance of the results presented here? I do not think so. The findings presented in this study on the socio-technical interactions between musicians and the platform and on the logics of cultural hierarchization on TikTok can be applied to various (and certainly core) topics of music-related research (see chapter 7). Additionally, the study provides an empirical basis for further research into music cultures in the platform context. Although it requires constant updating, such empirical research is urgently needed to better understand the *actual* cultural influence of digital platforms. The implementation of such projects is time-consuming. It is therefore important for academic researchers to be able to take this time, even when there is intense pressure to publish and even if the object of research is subject to rapid change.

Nevertheless, such processes necessitate a considerable degree of patience, not merely from the researchers themselves. The musicians who participated in this study also demonstrated remarkable patience, as they were required to wait a considerable amount of time before they could access the research findings. I wish to express my gratitude to all the interviewees, listed in alphabetical order below: Sam Ambers, Rachel Chiu, Stella Cole, Caity Gyorgy, Kellin Hanas, Erny Nunez, Brooklyn Stafford, and Stacey Ryan. The atmosphere in all the interviews was pleasant, the musicians were forthcoming and willing to provide information, and the conversations were illuminating. The interviewees offered me insights into the sphere of successful content creators, which is usually not accessible to researchers. I also found their videos, musicality, creativity, and humor appealing. Overall, I found working on this research project to be an enjoyable, educational, and stimulating experience.

It should be noted that my occasional critical assessments of TikTok should not be construed as a censure of the musicians who were interviewed or even of them personally. I am profoundly grateful to the people I interviewed for this study for their candidness and willingness to engage in discourse on challenging issues within the TikTok context. In this regard, the interviews proved to be particularly illuminating, and provided considerable insight into the logic of contemporary music and media cultures.

9. References

Abidin, Crystal. 2020. "Mapping Internet Celebrity on TikTok: Exploring Attention Economies and Visibility Labours." *Cultural Science* 12 (1): 77–103. https://doi.org/10.5334/csci.140.

Abidin, Crystal, and D. Bondy Valdovinos Kaye. 2021. "Audio Memes, Earworms, and Templatability: The 'Aural Turn' of Memes on TikTok." In *Critical Meme Reader: Global Mutations of the Viral Image*, edited by Chloë Arkenbout, Jack Wilson, and Daniel de Zeeuw, 58–68. Amsterdam: Institute of Network Cultures.

Ake, David, Charles Hiroshi Garrett, and Daniel Goldmark, eds. 2012. *Jazz/Not Jazz: The Music and Its Boundaries*. Berkeley, Los Angeles: University of California Press. https://doi.org/10.1525/california/9780520271036.001.0001.

Akrich, Madeleine. 1992. "The De-Scription of Technical Objects." In *Shaping Technology / Building Society: Studies in Sociotechnical Change*, edited by Wiebe E. Bijker and John Law, 205–24. Cambridge, London: MIT.

Ambers, Sam. 2021a. "feeling very good 🤪🤪." Accessed July 31, 2024. https://www.tiktok.com/@sam_ambers/video/6989743055075101957.

Ambers, Sam. 2021b. "taking applications for the cameraman role 😡😡." Accessed July 31, 2024. https://www.tiktok.com/@sam_ambers/video/6994495019105258758.

Ambers, Sam. 2021c. "wrong month, right time 👏👏🥵🥵." Accessed July 31, 2024. https://www.tiktok.com/@sam_ambers/video/6995948872023526661.

Ambers, Sam. 2022a. "an everyday occurrence." Accessed July 31, 2024. https://www.tiktok.com/@sam_ambers/video/7108775280390573317?lang=de-DE.

Ambers, Sam. 2022b. "Check out the lyrics whilst practising your performance with Multi Window on the Samsung Galaxy Z Flip4!" Accessed July 31, 2024. https://www.tiktok.com/@sam_ambers/video/7164322050922777862.

Apify. 2024. "TikTok Scraper." Accessed February 22, 2024. https://apify.com/clockworks/tiktok-scraper.

Appen, Ralf von, André Doehring, and Helmut Rösing. 2008. "Pop zwischen Historismus und Geschichtslosigkeit. Kanonbildungen in der populären Musik." In *No Time for Losers. Charts, Listen und andere Kanonisierungen in der populären Musik*, edited by Dietrich Helms and Thomas Phleps, 25–50. Bielefeld: transcript. https://doi.org/10.1515/9783839409831-003.

Arndt, Jürgen. 2006. "Nostalgische Schnittmuster im Jazz von Cassandra Wilson, Diana Krall und Norah Jones." In *Cut and paste: Schnittmuster populärer Musik der Gegenwart*, edited by Dietrich Helms and Thomas Phleps, 31–45. Bielefeld: transcript. https://doi.org/10.1515/9783839405697-003.

Arriagada, Arturo, and Francisco Ibáñez. 2020. "'You Need at Least One Picture Daily, If Not, You're Dead': Content Creators and Platform Evolution in the Social Media Ecology." *Social Media + Society* 6 (3): 1–12. https://doi.org/10.1177/2056305120944624.

Back, Andreas, and Peter Klose. 2018. "Jazzimprovisation zwischen spontaner Erfindung und erlernten Modellen. Ein empirischer Ansatz zur Analyse mit Hilfe von Video-Stimulated-Recall-Interviews." In *Pop weiter denken. Neue Anstöße aus Jazz Studies, Philosophie, Musiktheorie und Geschichte*, edited by Ralf von Appen and André Doehring, 31–52. Bielefeld: transcript. https://doi.org/10.1515/9783839446645-004.

Baker, Sarah, ed. 2015. *Preserving Popular Music Heritage. Do-It-Yourself, Do-It-Together*. London, New York: Routledge. https://doi.org/10.4324/9781315769882.

Baker, Sarah, ed. 2017. *Community Custodians of Popular Music's Past. A DIY Approach to Heritage*. London, New York: Routledge. https://doi.org/10.4324/9781315659923.

Banfield, William C. 2022. "It Don't Mean a Thing Without My Web Fan Base Thing: A Dance of Cultural Relevancy in Jazz Education Culture Today." In *Artistic Research in Jazz: Positions, Theories, Methods*, edited by Michael Kahr, 98–111. New York, Abingdon: Routledge. https://doi.org/10.4324/9780429275838-9.

Banks, Miranda, Bridget Conor, and Vicki Mayer, eds. 2016. *Production Studies, the Sequel! Cultural Studies of Global Media Industries*. New York, Abingdon: Routledge. https://doi.org/10.4324/9780429275838-9.

Barber, Simon. 2019. "My Jazz World: The Rise and Fall of a Digital Utopia." In *The Routledge Companion to Jazz Studies*, edited by Nicholas Gebhardt,

Nichole Rustin-Paschal, and Tony Whyton, 455–64. New York, Abingdon: Routledge. https://doi.org/10.4324/9781315315805-43.

Basch, Corey H., Bhavya Yalamanchili, and Joseph Fera. 2022. "#Climate Change on TikTok: A Content Analysis of Videos." *Journal of Community Health* 47: 163–67. https://doi.org/10.1007/s10900-021-01031-x.

Baym, Nancy, Rachel Bergmann, Raj Bhargava, Fernandez Diaz, Tarleton Gillespie, David Hesmondhalgh, Elena Maris, and Christopher J. Persaud. 2021. "Making Sense of Metrics in the Music Industries." *International Journal of Communication* 15: 3418–41.

Beer, David. 2013. *Popular Culture and New Media: The Politics of Circulation*. New York: Palgrave Macmillan. https://doi.org/10.1057/9781137270061.

Beer, David. 2017. "The Social Power of Algorithms." *Information, Communication & Society* 20 (1): 1–13. https://doi.org/10.1080/1369118X.2016.1216147.

Bell, Adam Patrick. 2015. "Can We Afford These Affordances? GarageBand and the Double-Edged Sword of the Digital Audio Workstation." *Action, Criticism & Theory for Music Education* 14 (1): 44–65.

Bell, Philip, and Marko Milic. 2002. "Goffman's Gender Advertisements Revisited: Combining Content Analysis with Semiotic Analysis." *Visual Communication* 1 (2): 203–22. https://doi.org/10.1177/147035720200100205.

Benjamin, Ruha. 2019. *Race After Technology: Abolitionist Tools for the New Jim Code*. Cambridge, Medford: Polity.

Bhandari, Aparajita, and Sara Bimo. 2022. "Why's Everyone on TikTok Now? The Algorithmized Self and the Future of Self-Making on Social Media." *Social Media + Society* 8 (1): 1–11. https://doi.org/10.1177/20563051221086241.

Biddle, Sam, Paulo Victor Ribeiro, and Tatiana Dias. 2020. "Invisible Censorship: TikTok Told Moderators to Suppress Posts by 'Ugly' People and the Poor to Attract New Users." *The Intercept*. Accessed February 21, 2024. https://theintercept.com/2020/03/16/tiktok-app-moderators-users-discrimination/.

Bishop, Sophie. 2018. "Anxiety, Panic and Self-Optimization: Inequalities and the YouTube Algorithm." *Convergence: The International Journal of Research into New Media Technologies* 24 (1): 69–84. https://doi.org/10.1177/1354856517736978.

Bishop, Sophie. 2019. "Managing Visibility on YouTube Through Algorithmic Gossip." *New Media & Society* 21 (11–12): 2589–2606. https://doi.org/10.1177/1461444819854731.

Bishop, Sophie. 2021. “Influencer Management Tools: Algorithmic Cultures, Brand Safety, and Bias.” *Social Media + Society* 7 (1): 1–13. https://doi.org/10.1177/20563051211003066.

Boffone, Trevor. 2022. “The D’Amelio Effect: TikTok, Charli D’Amelio, and the Construction of Whiteness.” In *TikTok Cultures in the United States*, edited by Trevor Boffone, 17–27. Abingdon, New York: Routledge. https://doi.org/10.4324/9781003280705-3.

Bogner, Alexander, Beate Littig, and Wolfgang Menz. 2014. *Interviews mit Experten. Eine praxisorientierte Einführung*. Wiesbaden: Springer VS. https://doi.org/10.1007/978-3-531-19416-5.

Bollenbeck, Georg. 2005. “Kulturkritik: Ein unterschätzter Reflexionsmodus der Moderne.” *Zeitschrift für Literaturwissenschaft und Linguistik* 137: 41–53. https://doi.org/10.1007/BF03379411.

Bonini, Tiziano, and Alessandro Gandini. 2019. “‘First Week Is Editorial, Second Week Is Algorithmic’: Platform Gatekeepers and the Platformization of Music Curation.” *Social Media + Society* 5 (4): 1–11. https://doi.org/10.1177/2056305119880006.

Brandellero, Amanda, and Susanne Janssen. 2014. “Popular Music as Cultural Heritage: Scoping Out the Field of Practice.” *International Journal of Heritage Studies* 20 (3): 224–40. https://doi.org/10.1080/13527258.2013.779294.

Bucher, Taina. 2012. “Want to Be on the Top? Algorithmic Power and the Threat of Invisibility on Facebook.” *New Media & Society* 14 (7): 1164–80. https://doi.org/10.1177/1461444812440159.

Bucher, Taina. 2018. *If...Then: Algorithmic Power and Politics*. New York: Oxford University Press. https://doi.org/10.1093/oso/9780190493028.001.0001.

Bucher, Taina, and Anne Helmond. 2018. “The Affordances of Social Media Platforms.” In *The SAGE Handbook of Social Media*, edited by Jean Burgess, Alice Marwick, and Thomas Poell, 233–53. London: SAGE. https://doi.org/10.4135/9781473984066.n14.

Burgess, Jean. 2006. “Hearing Ordinary Voices: Cultural Studies, Vernacular Creativity and Digital Storytelling.” *Continuum* 20 (2): 201–14. https://doi.org/10.1080/10304310600641737.

Burgess, Jean. 2021. “Platform Studies.” In *Creator Culture: An Introduction to Global Social Media Entertainment*, edited by Stuart Cunningham and David Craig, 21–38. New York: New York University Press. https://doi.org/10.18574/nyu/9781479890118.003.0005.

Burkey, Brant. 2020. "Repertoires of Remembering: A Conceptual Approach for Studying Memory Practices in the Digital Ecosystem." *Journal of Communication Inquiry* 44 (2): 178–97. https://doi.org/10.1177/0196859919852080.

Burkhart, Benjamin. 2019. "'Beschwingtes Deutschland?' Zur gegenwärtigen Jazzrezeption im deutschen Feuilleton." In *Jazzforschung heute. Themen, Methoden, Perspektiven*, edited by Martin Pfleiderer and Wolf-Georg Zaddach, 59–77. Berlin: Edition EMVAS.

Burkhart, Benjamin. 2022. "'Coffee Table Jazz': Curating Jazz on Spotify." Lecture, University of Music and Performing Arts, Graz, June 10, 2022.

Burkhart, Benjamin. 2024. "Zwischen den Stühlen: Zur (Nicht-)Erforschung von Jazz-Pop-Fusionen." In *"Parallelgesellschaften" in populärer Musik? Abgrenzungen – Annäherungen – Perspektiven*, edited by Ralf von Appen, Sarah Chaker, Michael Huber, and Sean Prieske, 213–37. Bielefeld: transcript. https://doi.org/10.1515/9783839473191-009.

Burkhart, Benjamin. 2025. "Jazz, Music Video, and Social Criticism: On Sons of Kemet's *In the Castle of My Skin*." In *Music Video and Transcultural Imaginaries. Media – Aesthetics – Social Utopia*, edited by Kathrin Dreckmann, Christofer Jost, and Bastian Schramm, 267–81. Münster, New York: Waxmann.

Caldwell, John Thornton. 2008. *Production Culture: Industrial Reflexivity and Critical Practice in Film and Television*. Durham: Duke University Press. https://doi.org/10.2307/j.ctv11sn72d.

Caldwell, John Thornton. 2013. "Zehn Thesen zur Produktionsforschung." *montage AV. Zeitschrift für Theorie und Geschichte audiovisueller Kommunikation* 22 (1): 33–47.

Chiu, Rachel. 2021a. "My Phone Was at 1% 🥴." Accessed July 31, 2024. https://www.tiktok.com/@rachelchiu1/video/6978547352948722949.

Chiu, Rachel. 2021b. "No Title." Accessed July 31, 2024. https://www.tiktok.com/@rachelchiu1/video/7020501035223436550.

Chiu, Rachel. 2021c. "Sorry My Pronunciation Is Trash I'm Canto." Accessed July 31, 2024. https://www.tiktok.com/@rachelchiu1/video/695368329080 09 34150.

Chiu, Rachel. 2022. "Singing While Trying to Do a Walking Bass Is NOT IT." Accessed July 31, 2024. https://www.tiktok.com/@rachelchiu1/video/7062 828071379946757.

Cho, Heekyoung. 2021. "The Platformization of Culture: Webtoon Platforms and Media Ecology in Korea and Beyond." *The Journal of Asian Studies* 80 (1): 73–93. https://doi.org/10.1017/S0021911820002405.

Cole, Stella. 2020. "Reply to @_amyjhs where my Julie London stans at??" Accessed July 31, 2024. https://www.tiktok.com/@stellakcole/video/6895187808378424581?lang=de-DE.

Collie, Natalie, and Caroline Wilson-Barnao. 2020. "Playing with TikTok: Algorithmic Culture and the Future of Creative Work." In *The Future of Creative Work: Creativity and Digital Disruption*, edited by Greg Hearn, 172–88. Cheltenham, Northampton: Edward Elgar. https://doi.org/10.4337/9781839101106.00020.

Collins, Patricia Hill, and Sirma Bilge. 2016. *Intersectionality*. Cambridge: Polity Press.

Cottell, Chris J. 2021. "Musical Videos as Works: Documenting Audiovisual Jazz Practices in Social Media Communities." *Sonic Scope: New Approaches to Audiovisual Culture* 2: 1–26. https://doi.org/10.21428/66f840a4.53ae3b48.

Cotter, Kelley. 2019. "Playing the Visibility Game: How Digital Influencers and Algorithms Negotiate Influence on Instagram." *New Media & Society* 21 (4): 895–913. https://doi.org/10.1177/1461444818815684.

Cotter, Kelley. 2022. "Practical Knowledge of Algorithms: The Case of Bread-Tube." *New Media & Society* 26 (4), 2131–50. https://doi.org/10.1177/14614448221081802.

Degele, Nina. 2006. "Bodification and Beautification: Zur Verkörperung sozialer und kultureller Differenzen durch Schönheitshandeln." In *Soziale Ungleichheit, kulturelle Unterschiede. Verhandlungen des 32. Kongresses der Deutschen Gesellschaft für Soziologie in München 2004. Teilband 1 und 2*, edited by Karl-Siegbert Rehberg, 579–92. Frankfurt am Main, New York: Campus.

Degele, Nina. 2017. "Schönheit und Attraktivität." In *Handbuch Körpersoziologie. Band 1: Grundbegriffe und Theoretische Perspektiven*, edited by Robert Gugutzer, Gabriele Klein, and Michael Meuser, 115–18. Wiesbaden: Springer VS. https://doi.org/10.1007/978-3-658-04136-6_19.

De la Motte-Haber, Helga. 2017. "Funktionale Musik." In *Handbuch Funktionale Musik: Psychologie – Technik – Anwendungsgebiete*, edited by Günther Rötter, 3–28. Wiesbaden: Springer. https://doi.org/10.1007/978-3-658-10219-7_1.

Dellisanti, Gabriele. 2022. "How Singer-Songwriter Laufey Is Bringing Jazz to the TikTok Generation." *Vogue Scandinavia*. Accessed February 22, 2024. https://www.voguescandinavia.com/articles/laufey.

DeVeaux, Scott. 1991. "Constructing the Jazz Tradition: Jazz Historiography." *Black American Literature Forum* 25 (3): 525–60. https://doi.org/10.2307/3041812.

DeVeaux, Scott, and Gary Giddins. 2015. *Jazz*. Second edition. New York: W.W. Norton.

Divon, Tom, and Tobias Ebbrecht-Hartmann. 2022. "#JewishTikTok: The JewToks' Fight Against Antisemitism." In *TikTok Cultures in the United States*, edited by Trevor Boffone, 47–58. Abingdon, New York: Routledge.

Dolata, Ulrich. 2013. *The Transformative Capacity of New Technologies: A Theory of Sociotechnical Change*. London, New York: Routledge. https://doi.org/10.4324/9780203070086.

Dolata, Ulrich. 2019. "Plattform-Regulierung. Koordination von Märkten und Kuratierung von Sozialität im Internet." *Berliner Journal für Soziologie* 29 (3–4): 179–206. https://doi.org/10.1007/s11609-020-00403-9.

Dolata, Ulrich. 2020a. "Internet – Platforms – Regulation. Coordination of Markets and Curation of Sociality: SOI Discussion Paper 2020–02." 1–29. Accessed July 31, 2024. https://www.sowi.uni-stuttgart.de/dokumente/forschung/soi/soi_2020_2_Dolata.Internet.Platforms.Regulation.pdf.

Dolata, Ulrich. 2020b. "The Digital Transformation of the Music Industry: The Second Decade: From Download to Streaming: SOI Discussion Paper 2020–04." 1–20. Accessed July 31, 2024. https://www.sowi.uni-stuttgart.de/dokumente/forschung/soi/soi_2020_4_Dolata.digital.transformation.music.industry.pdf.

Dolata, Ulrich. 2021. "Varieties of Internet Platforms and Their Transformative Capacity." In *The Future of Work*, edited by Christian Suter, Jacinto Cuvi, Philip Balsiger, and Mihaela Nedelcu, 100–16. Zurich, Geneva: Seismo.

Dolata, Ulrich, and Jan-Felix Schrape. 2023. "Platform Companies on the Internet as a New Organizational Form. A Sociological Perspective." *Innovation: The European Journal of Social Science Research* (online first): 1–20. https://doi.org/10.1080/13511610.2023.2182217.

Döring, Nicola. 2023. "Videoproduktion auf YouTube: Die Bedeutung von Geschlechterbildern." In *Handbuch Medien und Geschlecht. Perspektiven und Befunde der feministischen Kommunikations- und Medienforschung*, edited by Johanna Dorer, Brigitte Geiger, Brigitte Hipfl, and Viktorija Ratković, 963–73. Wiesbaden: Springer VS. https://doi.org/10.1007/978-3-658-20707-6_53.

Dreckmann, Kathrin, ed. 2021. *Musikvideo Reloaded: Über historische und aktuelle Bewegtbildästhetiken zwischen Pop, Kommerz und Kunst*. Düsseldorf: düsseldorf university press. https://doi.org/10.1515/9783110730623.

Dreckmann, Kathrin, and Elfi Vomberg, eds. 2023. *More Than Illustrated Music: Aesthetics of Hybrid Media Between Pop, Art and Video*. London: Bloomsbury. https://doi.org/10.5040/9781501381263.

Duffy, Brooke Erin, Thomas Poell, and David B. Nieborg, eds. 2019a. *Platform Practices in the Cultural Industries: Creativity, Labor, and Citizenship*: Special Issue of *Social Media + Society*.

Duffy, Brooke Erin, Thomas Poell, and David B. Nieborg. 2019b. "Platform Practices in the Cultural Industries: Creativity, Labor, and Citizenship." *Social Media + Society* 5 (4): 1–8. https://doi.org/10.1177/2056305119879672.

Duguay, Stefanie. 2018. "'The More I Look Like Justin Bieber in the Pictures, the Better': Queer Women's Self-Representation on Instagram." In *A Networked Self and Platforms, Stories, Connections*, edited by Zizi Papacharissi, 94–110. New York, Abingdon: Routledge. https://doi.org/10.4324/9781315193434-7.

Duguay, Stefanie. 2019. "'Running the Numbers': Modes of Microcelebrity Labor in Queer Women's Self-Representation on Instagram and Vine." *Social Media + Society* 5 (4): 1–11. https://doi.org/10.1177/2056305119894002.

Duguay, Stefanie, and Hannah Gold-Apel. 2023. "Stumbling Blocks and Alternative Paths: Reconsidering the Walkthrough Method for Analyzing Apps." *Social Media + Society* 9 (1): 1–10. https://doi.org/10.1177/20563051231158822.

Dunkel, Mario. 2014. "The Conceptualization of Race, Masculinity, and Authenticity in Early Jazz Historiography." In *Jazz, Gender, Authenticity: Proceedings of the 10th Nordic Jazz Research Conference, Stockholm August 30–31 2012*, edited by Alf Arvidsson, 25–34. Stockholm: Center for Swedish Folk Music and Jazz Research.

Eichhorn, Kate. 2013. *The Archival Turn in Feminism: Outrage in Order*. Philadelphia: Temple University Press.

Eidsheim, Nina Sun. 2019. *The Race of Sound: Listening, Timbre, and Vocality in African American Music*. Durham, London: Duke University Press. https://doi.org/10.1215/9781478090359.

Eisenegger, Mark. 2021. "Dritter, digitaler Strukturwandel der Öffentlichkeit als Folge der Plattformisierung." In *Digitaler Strukturwandel der Öffentlichkeit: Historische Verortung, Modelle und Konsequenzen*, edited by Mark Eisenegger, Marlis Prinzing, Patrik Ettinger, and Roger Blum, 17–39. Wiesbaden: Springer VS. https://doi.org/10.1007/978-3-658-32133-8_2.

Eriksson, Maria, Rasmus Fleischer, Anna Johansson, Pelle Snickars, and Patrick Vonderau. 2019. *Spotify Teardown: Inside the Black Box of Streaming Music*. Cambridge, London: MIT Press. https://doi.org/10.7551/mitpress/10932.001.0001.

Eriksson Krutrök, Moa. 2021. "Algorithmic Closeness in Mourning: Vernaculars of the Hashtag #grief on TikTok." *Social Media + Society* 7 (3): 1–12. https://doi.org/10.1177/20563051211042396.

Evans, Sandra K., Katy E. Pearce, Jessica Vitak, and Jeffrey W. Treem. 2017. "Explicating Affordances: A Conceptual Framework for Understanding Affordances in Communication Research." *Journal of Computer-Mediated Communication* 22 (1): 35–52. https://doi.org/10.1111/jcc4.12180.

Fitzgerald, Michael. 2012. "Reflecting on Jazz Archives: Past, Present, and Future." *Current Research in Jazz* 4. Accessed July 31, 2024. http://www.crj-online.org/v4/CRJ-ReflectingOnJazzArchives.php.

Flew, Terry. 2021. *Regulating Platforms*. Cambridge, Medford: Polity Press.

Foster, Jordan, and Jayne Baker. 2022. "Muscles, Makeup, and Femboys: Analyzing TikTok's 'Radical' Masculinities." *Social Media + Society* 8 (3): 1–14. https://doi.org/10.1177/20563051221126040.

Fox, Richard, Diamantis Panagiotopoulos, and Christina Tsouparopoulou. 2015. "Affordanz." In *Materiale Textkulturen. Konzepte – Materialien – Praktiken*, edited by Thomas Meier, Michael R. Ott, and Rebecca Sauer, 63–70. Berlin, Munich, Boston: De Gruyter. https://doi.org/10.1515/9783110371291.63.

Frith, Simon. 2007. "Is Jazz Popular Music?" *Jazz Research Journal* 1 (1): 7–23. https://doi.org/10.1558/jazz.v1i1.7.

Frith, Simon, and Simon Zagorski-Thomas, eds. 2012. *The Art of Record Production: An Introductory Reader for a New Academic Field*. Farnham: Ashgate.

Furia, Philip, and Laurie Patterson. 2016. *The American Song Book: The Tin Pan Alley Era*. New York: Oxford University Press.

Gabbard, Krin. 1995. "Introduction: The Jazz Canon and Its Consequences." In *Jazz Among the Discourses*, edited by Krin Gabbard, 1–28. Durham: Duke University Press. https://doi.org/10.1215/9780822397083-001.

Galloway, Kate. 2023. "Jazz Cats and Horse Licks: Animal Memes, Participatory Digital Culture, and Representing Jazz Online." *Jazz & Culture* 6 (2): 64–91. https://doi.org/10.5406/25784773.6.2.04.

Garofalo, Reebee, and Steve Waksman. 2013. *Rockin' Out: Popular Music in the U.S.A.* Sixth edition. Boston: Pearson.

Gebhardt, Nicholas, Nichole Rustin-Paschal, and Tony Whyton, eds. 2019. *The Routledge Companion to Jazz Studies*. New York, Abingdon: Routledge. https://doi.org/10.4324/9781315315805.

Geraghty, Lincoln. 2014. *Cult Collectors. Nostalgia, Fandom and Collecting Popular Culture*. London, New York: Routledge.

Gibbs, Martin, James Meese, Michael Arnold, Bjorn Nansen, and Marcus Carter. 2015. "#Funeral and Instagram: Death, Social Media, and Platform Vernacular." *Information, Communication & Society* 18 (3): 255–68. https://doi.org/10.1080/1369118X.2014.987152.

Gibson, James J. 1979. *The Ecological Approach to Visual Perception*. Boston: Houghton Mifflin Harcourt.

Gillespie, Tarleton. 2014. "The Relevance of Algorithms." In *Media Technologies: Essays on Communication, Materiality, and Society*, edited by Tarleton Gillespie, Pablo J. Boczkowski, and Kirsten A. Foot, 167–93. Cambridge, London: MIT Press.

Gillespie, Tarleton. 2017. "Algorithmically Recognizable: Santorum's Google Problem, and Google's Santorum Problem." *Information, Communication & Society* 20 (1): 63–80. https://doi.org/10.1080/1369118X.2016.1199721.

Gillespie, Tarleton. 2018. "Regulation of and by Platforms." In *The SAGE Handbook of Social Media*, edited by Jean Burgess, Alice Marwick, and Thomas Poell, 254–78. London: SAGE. https://doi.org/10.4135/9781473984066.n15.

Goldmann, Julia Elena, and Liesa Herbst. 2023. "Wer schön sein will ... Körpernormen und Schöhnheitsdiskurse in den Medien." In *Handbuch Medien und Geschlecht. Perspektiven und Befunde der feministischen Kommunikations- und Medienforschung*, edited by Johanna Dorer, Brigitte Geiger, Brigitte Hipfl, and Viktorija Ratković, 937–50. Wiesbaden: Springer VS. https://doi.org/10.1007/978-3-658-20707-6_72.

Gugutzer, Robert. 2022. *Soziologie des Körpers*. Sixth edition. Bielefeld: transcript. https://doi.org/10.36198/9783838557212.

Gyorgy, Caity. 2021a. "Can you tell which song I'm soloing on? Bonus points if you can count how many times I breathed!". Accessed July 31, 2024. https://www.tiktok.com/@caitygyorgy/video/7034321832908049669.

Gyorgy, Caity. 2021b. "Reply to @natashapenfield I'm so happy y'all like my originals:) here's another one!!💜." Accessed July 31, 2024. https://www.tiktok.com/@caitygyorgy/video/7024658201748458758.

Hanas, Kellin. 2021a. "bop dooba doo da bwiada women in jazz babyy 🤠🎺." Accessed July 31, 2024. https://www.tiktok.com/@kellinhanas/video/7018717730270383365?lang=de-DE.

Hanas, Kellin. 2021b. "depressive episodes? no. j a z z." Accessed July 31, 2024. https://www.tiktok.com/@kellinhanas/video/7015638515652381958?lang=de-DE.

Hautea, Samantha, Perry Parks, Bruno Takahashi, and Jing Zeng. 2021. "Showing They Care (Or Don't): Affective Publics and Ambivalent Climate Activism on TikTok." *Social Media + Society* 7 (2): 1–14. https://doi.org/10.1177/20563051211012344.

Hecken, Thomas. 2016. "Kulturkritik und Pop-Auffassungen. Definitionen und historische Ausprägungen der Kulturkritik." In *Kulturkritik und das Populä-*

re in der Musik, edited by Fernand Hörner, 17–32. Münster, New York: Waxmann.

Heile, Björn, Peter Elsdon, and Jenny Doctor, eds. 2016. *Watching Jazz: Encounters with Jazz Performance on Screen*. New York: Oxford University Press. https://doi.org/10.1093/acprof:oso/9780199347650.001.0001.

Helmond, Anne. 2015. "The Platformization of the Web: Making Web Data Platform Ready." *Social Media + Society* 1 (2): 1–11. https://doi.org/10.1177/2056305115603080.

Hesmondhalgh, David. 2022. "Streaming's Effects on Music Culture: Old Anxieties and New Simplifications." *Cultural Sociology* 16 (1): 3–24. https://doi.org/10.1177/17499755211019974.

Hjorth, Larissa, Heather Horst, Anne Galloway, and Genevieve Bell, eds. 2017. *The Routledge Companion to Digital Ethnography*. New York, Abingdon: Routledge. https://doi.org/10.4324/9781315673974.

Hoklas, Anne-Kathrin, and Steffen Lepa. 2015. "Mediales Musikhören im Alltag am Beispiel des Plattenspielers: Auditive Kultur aus der Perspektive der praxeologischen Wissenssoziologie." *Navigationen. Zeitschrift für Medien- und Kulturwissenschaften* 15 (2): 127–43.

Hopkins, Julian. 2020. "The Concept of Affordances in Digital Media." In *Handbuch soziale Praktiken und digitale Alltagswelten*, edited by Heidrun Friese, Gala Rebane, Marcus Nolden, and Miriam Schreiter, 47–54. Wiesbaden: Springer VS. https://doi.org/10.1007/978-3-658-08357-1_67.

Hurst, Alfred, and Philipp Mayring. 2005. "Qualitative Inhaltsanalyse." In *Qualitative Medienforschung. Ein Handbuch*, edited by Lothar Mikos and Claudia Wegener, 436–44. Konstanz: UVK.

Ile, Yvonne. 2021. "How Black Creators Stopped the Clock on TikTok." *INSAM Journal of Contemporary Music, Art and Technology* 8: 60–79. https://doi.org/10.51191/issn.2637-1898.2022.5.8.60.

Ilten, Carla. 2015. "'Use Your Skills to Solve This Challenge!' The Platform Affordances and Politics of Digital Microvolunteering." *Social Media + Society* 1 (2): 1–11. https://doi.org/10.1177/2056305115604175.

Jansson, André. 2023. "Beyond the Platform: Music Streaming as a Site of Logistical and Symbolic Struggle." *New Media & Society* 25 (12): 3203–21. https://doi.org/10.1177/14614448211036356.

Jenkins, Henry. 2006. *Convergence Culture: Where Old and New Media Collide*. New York, London: New York University Press.

Johnson, Mark R., and Jamie Woodcock. 2019. "'And Today's Top Donator Is': How Live Streamers on *Twitch.tv* Monetize and Gamify Their Broadcasts." *Social Media + Society* 5 (4): 1–11. https://doi.org/10.1177/2056305119881694.

Jones, Simon C., and Thomas G. Schumacher. 1992. "Muzak: On Functional Music and Power." *Critical Studies in Mass Communication* 9: 156–69. https://doi.org/10.1080/15295039209366822.

Jost, Christofer. 2017. "Professionalism as Style. Music Amateurs on YouTube and the Transformation of Production Techniques." In *Musik und Professionalität / Music and Professionalism*, edited by Knut Holtsträter and Michael Fischer, 55–69. Münster, New York: Waxmann.

Jost, Christofer. 2018. "Gedächtnisproduktion als webbasierte Aneignungspraxis. Populäre Songs und ihre Neuinterpretation auf Youtube." In *(Digitale) Medien und soziale Gedächtnisse*, edited by Gerd Sebald and Marie-Kristin Döbler, 83–104. Wiesbaden: Springer VS. https://doi.org/10.1007/978-3-658-19513-7_4.

Jost, Christofer. 2022. "Im Sog der Massenmedien. Professionalisiertes musikfilmisches Handeln von Jugendlichen auf YouTube." In *Jugend, Film, Musik*, edited by Kathrin Dreckmann, Carsten Heinze, Dagmar Hoffmann, and Dirk Matejovski, 407–26. Düsseldorf: düsseldorf university press.

Jost, Christofer, Axel Schmidt, and Klaus Neumann-Braun. 2014. "Innovation und Musik. Zur Transformation musikalischer Praxis im Horizont interaktionsmedialer Kommunikation." In *Handbuch Innovationen. Interdisziplinäre Grundlagen und Anwendungsfelder*, edited by Manfred Mai, 335–54. Wiesbaden: Springer VS. https://doi.org/10.1007/978-3-658-02317-1_18.

Jost, Ekkehard. 1998. "Der Jazz und die Politik. Szenen einer problematischen Beziehung." In *Populäre Musik, Politik und mehr... ein Forschungsmedley*, edited by Helmut Rösing and Thomas Phleps, 60–70. Karben: CODA.

Judd, Hannah. 2022. "Virals, Memes, and the Lick's Circulation Through Online Jazz Communities." *Twentieth-Century Music* 19 (3): 393–410. https://doi.org/10.1017/S1478572222000263.

Katel, Alisha. 2022. "Laufey Brings Jazz to Gen Z." *The Purbalite*. Accessed February 22, 2024. https://purbalite.net/25600/entertainment/laufey-brings-jazz-to-gen-z/.

Katz, Mark. 2010. *Capturing Sound: How Technology Has Changed Music*. Revised edition. Berkeley, Los Angeles: University of California Press. https://doi.org/10.1525/9780520947351.

Kaye, D. Bondy Valdovinos. 2022. "Please Duet This: Collaborative Music Making in Lockdown on *TikTok*." *Networking Knowledge* 15 (1): 59–77. https://doi.org/10.5406/25784773.6.2.05.

Kaye, D. Bondy Valdovinos. 2023. "JazzTok: Creativity, Community, and Improvisation on TikTok." *Jazz & Culture* 6 (2): 92–116.

Kaye, D. Bondy Valdovinos, Xu Chen, and Jing Zeng. 2021. "The Co-Evolution of Two Chinese Mobile Short Video Apps: Parallel Platformization of Douyin and TikTok." *Mobile Media & Communication* 9 (2): 229–53. https://doi.org/10.1177/2050157920952120.

Kaye, D. Bondy Valdovinos, Aleesha Rodriguez, Katrin Langton, and Patrik Wikström. 2021. "You Made This? I Made This: Practices of Authorship and (Mis)Attribution on TikTok." *International Journal of Communication* 15: 3195–215. https://doi.org/10.5210/spir.v2020i0.11354.

Kaye, D. Bondy Valdovinos, Jing Zeng, and Patrik Wikström. 2022. *TikTok: Creativity and Culture in Short Video*. Cambridge, Medford: Polity.

Keightley, Keir. 2008. "Music for Middlebrows: Defining the Easy Listening Era, 1946–1966." *American Music* 26 (3): 309–35. https://doi.org/10.2307/40071710.

Kennedy, Melanie. 2020. "'If the Rise of the TikTok Dance and E-Girl Aesthetic Has Taught Us Anything, It's That Teenage Girls Rule the Internet Right Now': TikTok Celebrity, Girls and the Coronavirus Crisis." *European Journal of Cultural Studies* 23 (6): 1069–76. https://doi.org/10.1177/1367549420945341.

Kim, Ji-Hyeon, and Jun Yu. 2019. "Platformizing Webtoons: The Impact on Creative and Digital Labor in South Korea." *Social Media + Society* 5 (4): 1–11. https://doi.org/10.1177/2056305119880174.

Klotz, Kelsey. 2023a. *Dave Brubeck and the Performance of Whiteness*. New York: Oxford University Press. https://doi.org/10.1093/oso/9780197525074.001.0001.

Klotz, Kelsey. 2023b. "Kenny G and the Ignorance of Whiteness." *Journal of Jazz Studies* 14 (1): 20–34. https://doi.org/10.14713/jjs.v14i1.248.

Klug, Daniel, Morgan Evans, and Geoff Kaufman. 2023. "How TikTok Served as a Platform for Young People to Share and Cope with Lived COVID-19 Experiences." *MedieKultur: Journal of Media and Communication Research* 38 (73): 152–70. https://doi.org/10.7146/mk.v38i73.128463.

Klug, Daniel, Morgan Evans, Yiluo Qin, and Geoff Kaufman. 2021. "Trick and Please. A Mixed-Method Study on User Assumptions About the TikTok Algorithm." *13th ACM Web Science Conference*, 84–92. https://doi.org/10.1145/3447535.3462512.

Knauer, Wolfram. 1990. *Zwischen Bebop und Free Jazz: Komposition und Improvisation des Modern Jazz Quartet*. Teilband 1: Text. Mainz: Schott.

Knauer, Wolfram, ed. 2016. *Gender and Identity in Jazz*. Hofheim: Wolke.

Knauer, Wolfram. 2018. "Was wäre denn politischer als die Utopie erschaffen? Jazz als musikalische Praxis der Gegenwart." *Musik & Ästhetik* 22 (86): 80–84.

Knauer, Wolfram. 2020. "Jazz und Politik – politischer Jazz? Eine bundesdeutsche Perspektive." In *Positionen! Jazz und Politik*, edited by Wolfram Knauer, 66–78. Hofheim: Wolke.

Korsgaard, Mathias Bonde. 2017. *Music Video After MTV: Audiovisual Studies, New Media, and Popular Music*. New York, Abingdon: Routledge. https://doi.org/10.4324/9781315617565.

Köver, Chris, and Markus Reuter. 2020. "Streit um TikTok: Im Zweifel mit der Partei." *Netzpolitik.org*. Accessed February 21, 2024. https://netzpolitik.org/2020/streit-um-tiktok-im-zweifel-mit-der-partei/.

Léveillé Gauvin, Hubert. 2018. "Drawing Listener Attention in Popular Music: Testing Five Musical Features Arising from the Theory of Attention Economy." *Musicae Scientiae* 22 (3): 291–304. https://doi.org/10.1177/1029864917698010.

Light, Ben, Jean Burgess, and Stefanie Duguay. 2018. "The Walkthrough Method: An Approach to the Study of Apps." *New Media & Society* 20 (3): 881–900. https://doi.org/10.1177/1461444816675438.

Lohmeier, Christine, Christian Schwarzenegger, and Maria Schreiber. 2020. "Instamemories. Geschichte in digitalen Medien als lebendige Erinnerungskultur jenseits formaler Bildungskontexte." *Medien + Erziehung* 20: 48–61. https://doi.org/10.21240/merz/2020.6.6.

Lucibello, Kristen M., Madison F. Vani, Alyona Koulanova, Melissa L. deJonge, Garcia Ashdown-Franks, and Catherine M. Sabiston. 2021. "#quarantine15: A Content Analysis of Instagram Posts During COVID-19." *Body Image* 38: 148–56. https://doi.org/10.1016/j.bodyim.2021.04.002.

Maasø, Arnt, and Anja Nylund Hagen. 2020. "Metrics and Decision-Making in Music Streaming." *Popular Communication* 18 (1): 18–31. https://doi.org/10.1080/15405702.2019.1701675.

Martinez, Wendyliz. 2022. "TikTok for Us by Us: Black Girlhood, Joy, and Self-Care." In *TikTok Cultures in the United States*, edited by Trevor Boffone, 39–46. Abingdon, New York: Routledge. https://doi.org/10.4324/9781003280705-5.

Mayer, Vicki, Miranda J. Banks, and John Thornton Caldwell, eds. 2009. *Production Studies: Cultural Studies of Media Industries*. New York, Abingdon: Routledge.

McDonald, Chris. 2006. "'Rock, Roll and Remember?': Addressing the Legacy of Jazz in Popular Music Studies." *Popular Music History* 1 (2): 125–45. https://doi.org/10.1558/pomh.v1i2.125.

McGee, Kristin. 2008. "The Feminization of Mass Culture and the Novelty of All-Girl Bands: The Case of the Ingenues." *Popular Music and Society* 31 (5): 629–62. https://doi.org/10.1080/03007760802188454.

McGee, Kristin. 2009. *Some Liked It Hot: Jazz Women in Film and Television, 1928–1959*. Middletown: Wesleyan University Press.

Medbøe, Haftor, and José Dias. 2014. "Improvisation in the Digital Age: New Narratives in Jazz Promotion and Dissemination." *First Monday* 19 (10). Accessed July 31, 2024. https://firstmonday.org/ojs/index.php/fm/article/view/5553/4132. https://doi.org/10.5210/fm.v19i10.5553.

Menke, Manuel, and Thomas Birkner. 2023. "Digitales Erinnern: Eine Synthese anhand der analytischen Kategorien Speichern, Datafizierung, Vernetzung und Affektivität." In *Handbuch Kommunikationswissenschaftliche Erinnerungsforschung*, edited by DFG-Netzwerk Kommunikationswissenschaftliche Erinnerungsforschung, 495–516. Berlin, Boston: De Gruyter. https://doi.org/10.1515/9783110629743-024.

Menke, Manuel, and Elke Grittmann. 2023. "Mediatisiertes Erinnern und seine Öffentlichkeiten." In *Handbuch Kommunikationswissenschaftliche Erinnerungsforschung*, edited by DFG-Netzwerk Kommunikationswissenschaftliche Erinnerungsforschung, 71–94. Berlin, Boston: De Gruyter. https://doi.org/10.1515/9783110629743-003.

Messmer, Roland. 2015. "Stimulated Recall als fokussierter Zugang zu Handlungs- und Denkprozessen von Lehrpersonen." *Forum Qualitative Sozialforschung / Forum: Qualitative Social Research* 16 (1): 1–20.

Michaelsen, René. 2013. "'The Song is Ended (but the Melody Lingers on).' Zu Kanonisierung und 'Standards' im Jazz." In *Der Kanon der Musik. Theorie und Geschichte. Ein Handbuch*, edited by Klaus Pietschmann and Melanie Wald-Fuhrmann, 626–48. München: Edition text + kritik.

Minadeo, Marisa, and Lizzy Pope. 2022. "Weight-Normative Messaging Predominates on TikTok: A Qualitative Content Analysis." *PLoS ONE* 17 (11): 1–12. https://doi.org/10.1371/journal.pone.0267997.

Moormann, Peter, Manuel Zahn, Patrick Bettinger, Sandra Hofhues, Helmke Jan Keden, and Kai Kaspar, eds. 2021. *Mikroformate. Interdisziplinäre Perspektiven auf aktuelle Phänomene in digitalen Medienkulturen*. München: kopaed.

Morris, Jeremy Wade. 2020. "Music Platforms and the Optimization of Culture." *Social Media + Society* 6 (3): 1–10. https://doi.org/10.1177/2056305120940690.

Morris, Jeremy Wade, Robert Prey, and David B. Nieborg. 2021. "Engineering Culture: Logics of Optimization in Music, Games, and Apps." *Review of Communication* 21 (2): 161–75. https://doi.org/10.1080/15358593.2021.1934522.

Nathaus, Klaus, and C. Clayton Childress. 2013. "The Production of Culture Perspective in Historical Research: Integrating the Production, Meaning and Reception of Symbolic Objects." *Zeithistorische Forschungen/Studies in Contemporary History* 10: 89–100.

Naujoks, Nick, Simone Weiß, and Marion Händel. 2021. "Lernstrategien beim musikalischen Üben handlungsnah erfassen: Ergebnisse einer Video-Stimulated Recall Interview Studie." *Beiträge empirischer Musikpädagogik* 12: 1–22. https://doi.org/10.62563/bem.v2021195.

Nguyen, Nga Thanh, Amanda McFadden, Donna Tangen, and Denise Beutel. 2013. "Video-Stimulated Recall Interviews in Qualitative Research." *Joint AARE Conference, Adelaide 2013*, 1–10.

Nieborg, David, and Thomas Poell. 2019. "The Platformization of Making Media." In *Making Media: Production, Practices, and Professions*, edited by Mark Deuze and Mirjam Prenger, 85–96. Amsterdam: Amsterdam University Press. https://doi.org/10.2307/j.ctvcj305r.8.

Nieborg, David B., Brooke Erin Duffy, and Thomas Poell, eds. 2020a. *Studying Platforms and Cultural Production: Methods, Institutions, and Practices*: Special Issue of *Social Media + Society*.

Nieborg, David B., Brooke Erin Duffy, and Thomas Poell. 2020b. "Studying Platforms and Cultural Production: Methods, Institutions, and Practices." *Social Media + Society* 6 (3): 1–7. https://doi.org/10.1177/2056305120943273.

Nieborg, David B., and Thomas Poell. 2018. "The Platformization of Cultural Production: Theorizing the Contingent Cultural Commodity." *New Media & Society* 20 (11): 4275–92. https://doi.org/10.1177/1461444818769694.

Noble, Safiya Umoja. 2018. *Algorithms of Oppression: How Search Engines Reinforce Racism*. New York: New York University Press.

Noonan, Ellen. 2012. *The Strange Career of Porgy and Bess: Race, Culture, and America's Most Famous Opera*. Chapel Hill: The University of North Carolina Press.

Nordgård, Daniel. 2021. "Views from the Boundaries: Music Streaming Revisited." *MedieKultur. Journal of Media and Communication Research* 70: 32–49. https://doi.org/10.7146/mediekultur.v37i70.122399.

Norman, Donald A. 1988. *The Psychology of Everyday Things*. New York: Basic Books.

Nunez, Erny. 2021. "up next is rags to riches!" Accessed July 31, 2024. https://www.tiktok.com/@young_crooner/video/6942617760140627206.

Paletschek, Sylvia, and Nina Reusch. 2013. "Populäre Geschichte und Geschlecht: Einleitung." In *Geschlecht und Geschichte in populären Medien*, edited by Elisabeth Cheauré, Sylvia Paletschek, and Nina Reusch, 7–37. Bielefeld: transcript. https://doi.org/10.14361/transcript.9783839423738.7.

Park, Seoyeon, Hyejin Jo, and Taeyoung Kim. 2023. "Platformization in Local Cultural Production: Korean Platform Companies and the K-Pop Industry." *International Journal of Communication* 17: 2422–43.

Pellegrinelli, Lara. 2008. "Separated at 'Birth': Singing and the History of Jazz." In *Big Ears: Listening for Gender in Jazz Studies*, edited by Kathryn Ryan and Sherrie Tucker, 32–47. Durham, London: Duke University Press. https://doi.org/10.2307/j.ctv1134ftp.5.

Pelly, Liz. 2017. "The Problem with Muzak: Spotify's Bid to Remodel an Industry." *The Baffler*. Accessed July 31, 2024. https://thebaffler.com/salvos/the-problem-with-muzak-pelly.

Pelly, Liz. 2018. "Streambait Pop: The Emergence of a Total Spotify Genre." *The Baffler*. Accessed July 31, 2024. https://thebaffler.com/downstream/streambait-pop-pelly.

Pentzold, Christian. 2021. "Digitale Gedächtnisse." In *Handbuch sozialwissenschaftliche Gedächtnisforschung*, edited by Mathias Berek, Kristina Chmelar, Oliver Dimbath, Hanna Haag, Michael Heinlein, Nina Leonhard, Valentin Rauer, and Gerd Sebald, 1–15. Wiesbaden: Springer VS.

Pentzold, Christian, Christine Lohmeier, and Anne Kaun. 2023. "Was will, was kann, was soll eine kommunikationswissenschaftliche Erinnerungsforschung?" In *Handbuch Kommunikationswissenschaftliche Erinnerungsforschung*, edited by DFG-Netzwerk Kommunikationswissenschaftliche Erinnerungsforschung, 1–43. Berlin, Boston: De Gruyter. https://doi.org/10.1515/9783110629743-001.

Peress, Maurice. 2004. *Dvořák to Duke Ellington: A Conductor Explores America's Music and Its African American Roots*. Oxford, New York: Oxford University Press. https://doi.org/10.1093/acprof:oso/9780195098228.001.0001.

Peterson-Salahuddin, Chelsea. 2024. "'Pose': Examining Moments of 'Digital' Dark Sousveillance on TikTok." *New Media & Society* 26 (5): 2387–406. https://doi.org/10.1177/14614448221080480.

Pillai, Nicolas. 2018. "A Star Named Miles. Tracking Jazz Musicians Across Media." In *Jazz @ 100: An Alternative to a Story of Heroes*, edited by Wolfram Knauer, 267–76. Hofheim: Wolke.

Pink, Sarah, Heather A. Horst, John Postill, Larissa Hjorth, Tania Lewis, and Jo Tacchi. 2016. *Digital Ethnography: Principles and Practice*. Los Angeles: SAGE.

Plantin, Jean-Christophe, Carl Lagoze, Paul N. Edwards, and Christian Sandvig. 2018. "Infrastructure Studies Meet Platform Studies in the Age of Google and Facebook." *New Media & Society* 20 (1): 293–310. https://doi.org/10.1177/1461444816661553.

Poell, Thomas, David Nieborg, and Brooke Erin Duffy. 2022. *Platforms and Cultural Production*. Cambridge, Medford: Polity.

Pomerantz, Shauna, and Miriam Field. 2022. "Watching TikTok, Feeling Feminism." In *TikTok Cultures in the United States*, edited by Trevor Boffone, 61–71. Abingdon, New York: Routledge. https://doi.org/10.4324/9781003280705-8.

Prey, Robert. "Locating Power in Platformization: Music Streaming Playlists and Curatorial Power." *Social Media + Society* 6 (2): 1–11. https://doi.org/10.1177/2056305120933291.

Prouty, Ken. 2012a. "Creating Boundaries in the Virtual Jazz Community." In *Jazz/Not Jazz: The Music and Its Boundaries*, edited by David Ake, Charles H. Garrett, and Daniel Goldmark, 70–88. Berkeley, Los Angeles: University of California Press. https://doi.org/10.1525/california/9780520271036.003.0005.

Prouty, Ken. 2012b. *Knowing Jazz: Community, Pedagogy, and Canon in the Information Age*. Jackson: University Press of Mississippi. https://doi.org/10.14325/mississippi/9781617031632.001.0001.

Przyborski, Aglaja, and Monika Wohlrab-Sahr. 2021. *Qualitative Sozialforschung. Ein Arbeitsbuch*. Fifth and revised edition. Berlin, Boston: De Gruyter. https://doi.org/10.1515/9783110710663.

Radovanović, Bojana. 2022. "TikTok and Sound: Changing the Ways of Creating, Promoting, Distributing and Listening to Music." *INSAM Journal of Contemporary Music, Art and Technology* 9: 51–73. https://doi.org/10.51191/issn.2637-1898.2022.5.9.51.

Raffa, Massimiliano, and Riccardo Pronzato. 2021. "The Algorithmic Imaginary of Cultural Producers. Towards Platform-Optimized Music?" *H-ermes. Journal of Communication* 19: 293–321.

Reddan, James, Monika Herzig, and Michael Kahr, eds. 2022. *The Routledge Companion to Jazz and Gender*. New York, Abingdon: Routledge. https://doi.org/10.4324/9781003081876.

Reitsamer, Rosa. 2018. "Gendered Narratives of Popular Music History and Heritage." In *The Routledge Companion to Popular Music History and Heritage*, edited by Sarah Baker, Catherine Strong, Lauren Istvandity, and Zelmarie Cantillon, 26–35. London, New York: Routledge. https://doi.org/10.4324/9781315299310-3.

Reynolds, Dean S. 2017. *Jazz and Recording in the Digital Age: Technology, New Media, and Performance in New York and Online:* Dissertation: The City University of New York. Accessed July 31, 2024. https://academicworks.cuny.edu/gc_etds/2404/.

Roberge, Jonathan, and Robert Seyfert. 2018. "What Are Algorithmic Cultures?" In *Algorithmic Cultures: Essays on Meaning, Performance and New Technologies*, edited by Robert Seyfert and Jonathan Roberge, 1–25. Abingdon, New York: Routledge.

Rochford, Elle, and Zachary D. Palmer. 2022. "Trans TikTok: Sharing Information and Forming Community." In *TikTok Cultures in the United States*, edited by Trevor Boffone, 84–94. Abingdon, New York: Routledge. https://doi.org/10.4324/9781003280705-10.

Rustin, Nichole T., and Sherrie Tucker, eds. 2008. *Big Ears: Listening for Gender in Jazz Studies*. Durham, London: Duke University Press. https://doi.org/10.1215/9780822389224.

Ryan, Kathryn, and Antoaneta Tileva. 2022. "Taking the Past Out of the Pastoral: TikTok's Queer 'Cottagecore' Culture and Performative Placemaking." *Queer Studies in Media & Popular Culture* 7 (3): 165–76. https://doi.org/10.1386/qsmpc_00077_1.

Ryan, Stacey. 2020. "a bunch of people said my wap video sounded like the Andrew Sisters so here's Boogie Woogie Bugle Boy!". Accessed July 31, 2024. https://www.tiktok.com/@staceyryanmusic/video/6873273255545883905.

Ryan, Stacey. 2021a. "#duet with @ubiran this what happens when jazz meets window (credit to @carlosxcordova he did it first!)." Accessed July 31, 2024. https://www.tiktok.com/@staceyryanmusic/video/6920325211267124486.

Ryan, Stacey. 2021b. "Reply to @_im_yaboy you REALLY liked this original so now I want to see some COOL #DUETS." Accessed July 31,

2024. https://www.tiktok.com/@staceyryanmusic/video/7047292169790541062?lang=de-DE.

Sadler, Olivia. 2022. "Defiant Amplification or Decontextualized Commercialization? Protest Music, TikTok, and Social Movements." *Social Media + Society* 8 (2): 1–10. https://doi.org/10.1177/20563051221094769.

Savic, Milovan. 2021. "From Musical.ly to TikTok: Social Construction of 2020's Most Downloaded Short-Video App." *International Journal of Communication* 15: 3173–94.

Schellewald, Andreas. 2021. "Communicative Forms on TikTok: Perspectives from Digital Ethnography." *International Journal of Communication* 15: 1437–57.

Schrape, Jan-Felix. 2021. *Digitale Transformation*. Bielefeld: transcript. https://doi.org/10.36198/9783838555805.

Schuegraf, Martina. 2023. "Hollywood-Stars und Social Media Celebrities." In *Handbuch Medien und Geschlecht. Perspektiven und Befunde der feministischen Kommunikations- und Medienforschung*, edited by Johanna Dorer, Brigitte Geiger, Brigitte Hipfl, and Viktorija Ratković, 951–61. Wiesbaden: Springer VS. https://doi.org/10.1007/978-3-658-20707-6_54.

Scolere, Leah, Urszula Pruchniewska, and Brooke Erin Duffy. 2018. "Constructing the Platform-Specific Self-Brand: The Labor of Social Media Promotion." *Social Media + Society* 4 (3): 1–11. https://doi.org/10.1177/2056305118784768.

Seabrook, John. 2015. *The Song Machine: Inside the Hit Factory*. New York: W.W. Norton.

Seaver, Nick. 2022. *Computing Taste: Algorithms and the Makers of Music Recommendation*. Chicago, London: Chicago University Press. https://doi.org/10.7208/chicago/9780226822969.001.0001.

Slack, Jennifer Daryl, and Stefka Hristova. 2020. "Why We Need the Concept of Algorithmic Culture." In *Algorithmic Culture: How Big Data and Artificial Intelligence Are Transforming Everyday Life*, edited by Stefka Hristova, Soonkwan Hong, and Jennifer D. Slack, 15–33. Lanham: Lexington. https://doi.org/10.5771/9781793635747-15.

Smit, Rik. 2023. "Die Plattformisierung des Erinnerns." In *Handbuch Kommunikationswissenschaftliche Erinnerungsforschung*, edited by DFG-Netzwerk Kommunikationswissenschaftliche Erinnerungsforschung, 471–93. Berlin, Boston: De Gruyter. https://doi.org/10.1515/9783110629743-023.

Smudits, Alfred. 2007. "Medien, Mediamorphosen und Jazz." In *Festschrift Franz Kerschbaumer zum 60. Geburtstag*, edited by Franz Krieger and Bernd Hoffmann, 193–202. Graz: Akademische Druck- und Verlagsanstalt.

Solis, Gabriel. 2019. "Soul, Afrofuturism & the Timeliness of Contemporary Jazz Fusions:" *Daedalus* 148 (2): 23–35. https://doi.org/10.1162/daed_a_01740.

Southerton, Clare. 2021. "Lip-Syncing and Saving Lives: Healthcare Workers on TikTok." *International Journal of Communication* 15: 3248–68.

Spencer, Scott B. 2023. "Meme and Variations: How Video Mashups of John Coltrane's Giant Steps Became a Thing." In *Remediating Sound: Repeatable Culture, YouTube and Music*, edited by Holly Rogers, Joana Freitas, and João F. Porfírio, 223–42. New York, London, Dublin: Bloomsbury. https://doi.org/10.5040/9781501387357.ch-010.

Stafford, Brooklyn. 2021a. "Here's my take on this piece 🥰🎶." Accessed July 31, 2024. https://www.tiktok.com/@brooklyn.stafford/video/7021150775162178862.

Stafford, Brooklyn. 2021b. "The mashup you didn't ask for but needed 🥴🎶." Accessed July 31, 2024. https://www.tiktok.com/@brooklyn.stafford/video/7005622424368581894.

Stafford, Brooklyn. 2022. "What would Mozart say?😩🎹." Accessed July 31, 2024. https://www.tiktok.com/@brooklyn.stafford/video/7067945616399273262.

Steinke, Ines. 2017. "Gütekriterien qualitativer Forschung." In *Qualitative Forschung. Ein Handbuch*. Twelfth edition, edited by Uwe Flick, Ernst von Kardorff, and Ines Steinke, 319–31. Reinbek bei Hamburg: Rowohlt.

Sterne, Jonathan. 1997. "Sounds Like the Mall of America: Programmed Music and the Architectonics of Commercial Space." *Ethnomusicology* 41 (1): 22–50. https://doi.org/10.2307/852577.

Stoever, Jennifer Lynn. 2016. *The Sonic Color Line: Race and the Cultural Politics of Listening*. New York: New York University Press.

Striphas, Ted. 2015. "Algorithmic Culture." *European Journal of Cultural Studies* 18 (4–5): 395–412. https://doi.org/10.1177/1367549415577392.

Strübing, Jörg. 2014. *Grounded Theory. Zur sozialtheoretischen und epistemologischen Fundierung eines pragmatistischen Forschungsstils*. Third edition. Wiesbaden: Springer VS. https://doi.org/10.1007/978-3-531-19897-2.

Strübing, Jörg. 2022. "Grounded Theory und Theoretical Sampling." In *Handbuch Methoden der Empirischen Sozialforschung*, edited by Nina Baur and Jörg

Blasius, 587–606. Wiesbaden: Springer VS. https://doi.org/10.1007/978-3-658-37985-8_37.

Sundet, Vilde Schanke. 2021. *Television Drama in the Age of Streaming: Transnational Strategies and Digital Production Cultures at the NRK*. London: Palgrave Macmillan. https://doi.org/10.1007/978-3-030-66418-3.

Suzuki, Yoko. 2013. "Two Strikes and the Double Negative: The Intersections of Gender and Race in the Cases of Female Jazz Saxophonists." *Black Music Research Journal* 33 (2): 207–26. https://doi.org/10.5406/blacmusiresej.33.2.0207.

Sweeney-Romero, Katlin Marisol. 2022. "Wellness TikTok: Morning Routines, Eating Well, and Getting Ready to Be 'That Girl'." In *TikTok Cultures in the United States*, edited by Trevor Boffone, 108–16. Abingdon, New York: Routledge. https://doi.org/10.4324/9781003280705-13.

Sykes, Tom. 2009. "Remediated Swing and the Celestial Jukebox: New 'Friends' for Jazz?" In *SPARC 2009 Proceedings*, edited by Victoria Sheppard, 390–95.

Sykes, Tom. 2012. "Virtual Scenes of Interaction: Jazz and Connectedness in the Digital Age." *Jazzforschung / Jazz Research* 44: 75–89.

Sykes, Tom. 2013. *Jazz for the iPod Generation: Digital Media and Jazz in the UK*. Dissertation. University of Salford. Accessed July 31, 2024. https://salford-repository.worktribe.com/output/1427907/jazz-for-the-ipod-generation-digital-media-and-jazz-in-the-uk.

Sykes, Tom. 2017. "Making Scenes: Social Media and New Conceptions of Jazz Communities." In *New Jazz Conceptions: History, Theory, Practice*, edited by Roger Fagge and Nicolas Pillai, 36–62. Abingdon, New York: Routledge.

Tackley, Catherine. 2012. *Benny Goodman's Famous 1938 Carnegie Hall Jazz Concert*. Oxford, New York: Oxford University Press. https://doi.org/10.1093/acprof:osobl/9780195398304.001.0001.

TikTok. n. d. "TikTok Creator Fund Terms." Accessed February 27, 2024. https://www.tiktok.com/legal/page/global/tiktok-creator-fund-terms/en.

TikTok. 2021a. "The TikTok Creator Fund Is Now LIVE in the UK, Germany, Italy, France and Spain, and Here Is How to Apply!". Accessed February 27, 2024. https://newsroom.tiktok.com/en-gb/the-tiktok-creator-fund-is-now-live-across-europe-and-here-is-how-to-apply.

TikTok. 2021b. "TikTok Creator Fund: Your Questions Answered." Accessed February 21, 2024. https://newsroom.tiktok.com/en-gb/tiktok-creator-fund-your-questions-answered.

Tomasena, José M. 2019. "Negotiating Collaborations: BookTubers, the Publishing Industry, and YouTube's Ecosystem." *Social Media + Society* 5 (4): 1–12. https://doi.org/10.1177/2056305119894004.

Tucker, Sherrie. 2000. *Swing Shift: "All-Girl" Bands of the 1940s*. Durham, London: Duke University Press. https://doi.org/10.2307/j.ctv11smw66.

Tucker, Sherrie. 2002. "Women." In *The New Grove Dictionary of Jazz. Second Edition. Volume Three: Nightclubs–Zwingenberger*, edited by Barry Kernfeld, 978–84. Oxford: Oxford University Press.

Van der Merwe, Ann. 2017. *The American Songbook: Music for the Masses*. Lanham: Rowman & Littlefield.

Van Dijck, José, Thomas Poell, and Martijn de Waal. 2018. *The Platform Society: Public Values in a Connective World*. New York: Oxford University Press. https://doi.org/10.1093/oso/9780190889760.001.0001.

Vernallis, Carol. 2013. *Unruly Media: YouTube, Music Video, and the New Digital Cinema*. Oxford, New York: Oxford University Press. https://doi.org/10.1093/acprof:oso/9780199766994.001.0001.

Vonderau, Patrick. 2013. "Theorien zur Produktion: Ein Überblick." *montage AV. Zeitschrift für Theorie und Geschichte audiovisueller Kommunikation* 22 (1): 9–32.

Vonderau, Patrick. 2023. "What Is Production Studies Anyway?" In *Produktionskulturen audiovisueller Medien. Neuere Perspektiven der Medienindustrie- und Produktionsforschung*, edited by Stefan Udelhofen, Dennis Göttel, and Aycha Riffi, 33–52. Wiesbaden: Springer VS. https://doi.org/10.1007/978-3-658-38412-8_3.

Wall, Tim, and Andrew Dubber. 2010. "Experimenting with Fandom, Live Music, and the Internet: Applying Insights from Music Fan Culture to New Media Production." *Journal of New Music Research* 39 (2): 159–69. https://doi.org/10.1080/09298215.2010.489645.

Washburne, Christopher. 2004. "Does Kenny G Play Bad Jazz? A Case Study." In *Bad Music: The Music We Love to Hate*, edited by Christopher Washburne and Maiken Derno, 123–47. New York, Abingdon: Routledge.

West, Aaron J. 2008. *Caught Between Jazz and Pop: The Contested Origins, Criticism, Performance Practice, and Reception of Smooth Jazz*. Dissertation: University of North Texas. Accessed July 31, 2024. https://digital.library.unt.edu/ark:/67531/metadc9722/.

Whyton, Tony. 2010. *Jazz Icons: Heroes, Myths and the Jazz Tradition*. Cambridge, New York: Cambridge University Press.

Whyton, Tony. 2013. *Beyond* A Love Supreme*: John Coltrane and the Legacy of an Album*. Oxford, New York: Oxford University Press. https://doi.org/10.1093/acprof:osobl/9780199733231.001.0001.

Williams, Katherine. 2023. "Jazz and the American Songbook." In *Jazz and American Culture*, edited by Michael Borshuk, 262–73. Cambridge, New York: Cambridge University Press. https://doi.org/10.1017/9781009420167.022.

Williams, Sophie. 2022. "Homeward Bound: The Night Laufey Became an Icelandic National Treasure." *New Musical Express*. Accessed February 22, 2024. https://www.nme.com/features/music-interviews/laufey-interview-jazz-debut-album-iceland-radar-3343047.

Winker, Gabriele, and Nina Degele. 2009. *Intersektionalität. Zur Analyse sozialer Ungleichheiten*. Bielefeld: transcript. https://doi.org/10.1515/9783839411490.

Wright, Brian F. 2023. "Introduction: What Kenny G Can Teach Us About Jazz." *Journal of Jazz Studies* 14 (1): 1–19. https://doi.org/10.14713/jjs.v14i1.247.

Zagorski-Thomas, Simon. 2014. *The Musicology of Record Production*. Cambridge: Cambridge University Press. https://doi.org/10.1017/CBO9781139871846.

Zagorski-Thomas, Simon, Katia Isakoff, Sophie Stévance, and Serge Lacasse, eds. 2019. *The Art of Record Production, Volume 2: Creative Practice in the Studio*. Abingdon, New York: Routledge. https://doi.org/10.4324/9781315467658.

Zappavigna, Michele. 2016. "Social Media Photography: Construing Subjectivity in Instagram Images." *Visual Communication* 15 (3): 271–92. https://doi.org/10.1177/1470357216643220.

Zeng, Jing, and Crystal Abidin. 2021. "'#OkBoomer, Time to Meet the Zoomers': Studying the Memefication of Intergenerational Politics on TikTok." *Information, Communication & Society* 24 (16): 2459–81. https://doi.org/10.1080/1369118X.2021.1961007.

Zeng, Jing, Crystal Abidin, and Mike S. Schäfer. 2021. "Research Perspectives on TikTok and Its Legacy Apps." *International Journal of Communication* 15: 3161–72.

Zeng, Jing, and D. Bondy Valdovinos Kaye. 2022. "From Content Moderation to Visibility Moderation: A Case Study of Platform Governance on TikTok." *Policy & Internet* 14: 79–95. https://doi.org/10.1002/poi3.287.

Zeng, Jing, Mike S. Schäfer, and Joachim Allgaier. 2021. "Reposting 'Till Albert Einstein Is TikTok Famous': The Memetic Construction of Science on TikTok." *International Journal of Communication* 15: 3216–47. https://doi.org/10.31219/osf.io/8tdvm.

Zillien, Nicole. 2008. “Die (Wieder-)Entdeckung der Medien. Das Affordanzkonzept in der Mediensoziologie.” *Sociologia Internationalis* 46 (2): 161–81. https://doi.org/10.3790/sint.46.2.161.

Zulli, Diana, and David James Zulli. 2022. “Extending the Internet Meme: Conceptualizing Technological Mimesis and Imitation Publics on the TikTok Platform.” *New Media & Society* 24 (8): 1872–90. https://doi.org/10.1177/1461444820983603.